Data Visualization
in the Geological Sciences

JAMES R. CARR

UNIVERSITY OF NEVADA–RENO
MACKAY SCHOOL OF MINES

Data Visualization
in the Geological Sciences

PRENTICE HALL
Upper Saddle River, NJ 07458

Library of Congress Cataloging-in-Publication Data

Carr, James R.
    Data visualization in the geological sciences/James R. Carr.
        p. cm.
    Includes bibliographical references and index.
    ISBN 0-13-089706-X
        1. Geology—Statisical methods. 2. Geology—Statisical Methods—Data processing.
        3. Geology—Graphic methods. 4. Geology—Graphic methods—Data processing.   I. Title.

QE33.2.S82 C37 2002
551'.0285'66—dc21

                              2001052353

Senior Editor: *Patrick Lynch*
Marketing Manager: *Christine Henry*
Assistant Managing Editor: *Beth Sturla*
Production/Composition: *WestWords, Inc.*
Manufacturing Manager: *Trudy Pisciotti*
Assistant Manufacturing Manager: *Michael Bell*
Assistant Managing Editor, Science Media: *Nicole Bush*
Media Production Editor: *Elizabeth Wright*
Art Director: *Jayne Conte*
Cover Designer: *Bruce Kenselaar*
Managing Editor, Audio/Visual Assets: *Grace Hazeldine*
Art Editor: *Adam Velthaus*
Assistant Editor: *Amanda Griffith*
Editorial Assistant: *Sean Hale*

 © 2002 by Prentice-Hall, Inc.
Upper Saddle River, New Jersey 07458

Printed in the United States of America
10 9 8 7 6 5 4 3 2 1

ISBN 0-13-089706-X

Pearson Education Ltd., *London*
Pearson Education Australia Pty. Limited, *Sydney*
Pearson Education *Singapore*, Pte. Ltd.
Pearson Education North Asia Ltd., *Hong Kong*
Pearson Education Canada, Ltd., *Toronto*
Pearson Educatión de Mexico, S.A. de C.V.
Pearson Education—Japan, *Tokyo*
Pearson Education Malaysia, Pte. Ltd.

For

Elwood Messick (Russell E. Carr), San Francisco, California, 1926–1996
Elsie E. (Hersley) Carr, Red Deer, Alberta, Canada, 1926–
Ivy E. Freeman, Yerington, Nevada, 1916–1991
Anna E. (Kellison) Freeman, Reno, Nevada, 1918–1997

Thank you for the gift of life

CONTENTS

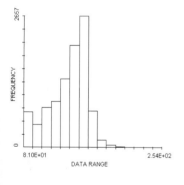

CHAPTER 3

Bivariate Data Analysis

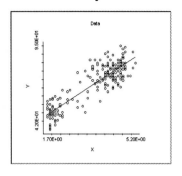

CHAPTER 4

Multivariate Data Analysis

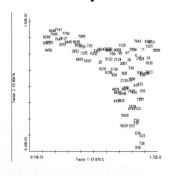

CHAPTER 5   85

Univariate Spatial Analysis

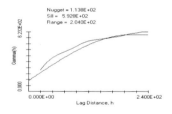

CHAPTER 6   132

Multivariate Spatial Data Analysis

CHAPTER 7   159

Spatial Simulation

CHAPTER 8   180

Digital Image Processing

CHAPTER 9   219

Composite Visualizations

PREFACE

This book is designed for any course, undergraduate or graduate, that involves the analysis of data useful to the study of geology, geography, agricultural sciences such as forestry, rangelands, and soil science, environmental science, ecology, civil engineering, geological engineering, environmental engineering and mining engineering. This book emphasizes applications of data analysis in these fields. The emphasis is on visual interpretation of data and of results from their analysis.

Readers will learn about statistics that describe data. From this fundamental foundation, more sophisticated algorithms for analyzing and manipulating data are applied. First, regression analysis is described and shown to yield a metric for correlation. This metric is of paramount importance to multivariate analysis. Moreover, space is of critical concern to many scientific and engineering applications, consequently applications in spatial analysis, both univariate and multivariate, are described in which correlation is, again, the metric of paramount importance. Spatial simulations are then shown to yield visually exciting images that enable experiments with spatial data to enhance the understanding of their behavior. A widely utilized form of spatial data is that which is obtained from digital sensors, spaceborne sensors in particular. Methods of data analysis described in the text are applied to these data to yield visualizations that aid a reader's understanding of them. Finally, composite visualizations are introduced, such as the draping of Landsat TM data over digital elevation to understand how image features correlate with elevation. Such composite visualizations are also used to associate color with correlation. In general, the unifying theme of this text is the analysis and understanding of correlation.

Following an introductory chapter that motivates the use and understanding of the text, chapters are presented in order on univariate data analysis, bivariate data analysis (regression and correlation), multivariate data analysis, univariate spatial analysis, multivariate spatial analysis, spatial simulation (including experiments with random number generation), digital image processing, and composite visualizations. A CD is included at the back of the text that presents a program, *Visual_Data*, a Windows-based program capable of reproducing all analytical results shown in the text. Moreover, the CD contains all data sets and digital images shown in the text, as well as many that are not shown to enable readers to experiment with data analysis beyond what is shown in the text. Readers are not forced to use *Visual_Data*. Instead, this program is provided should other software not be available. The text presents many examples of data analysis using *Microsoft's Excel* spreadsheet software and MATLAB, two powerful, commercial software platforms should readers prefer their use.

This book is written from the perspective of eighteen years of college-level instruction on data analysis in geological engineering and geology programs. This experience has shown that too much emphasis on mathematics when describing algorithms leaves too little time for a discussion of applications. But, a superficial treatment of mathematics trivializes algorithms, risking their use as "black boxes." This text was written to attempt a balance between the mathematical description of algorithms and visualizations of their outcomes. Moreover, these visualizations provide useful insights to algorithms that are not necessarily obvious in their equation form. Most importantly, visualizations of data are emphasized. A single data set, a collection of Landsat TM spectral bands, is used throughout this text as a unifying theme, showing how seemingly independent algorithms for data analysis are, in fact, interrelated, moreover showing that useful analyses of data can be obtained even if they don't conform to some ideal notion of statistical analysis, such as conformity to the normal distribution. A fundamental aspect of this text is on visualizing the data themselves, thinking critically of all outcomes from their analysis to judge whether useful insight to their behavior has been achieved.

Data Visualization
in the Geological Sciences

Introduction

When your mind can truly see the picture that a few numbers paint, the pure essence of mathematics is realized.

1.1 Purpose

Visualization is herein defined as the creation of graphical images directly from data, their transforms, and outcomes from algorithmic manipulation of these data. Mathematical algorithms are necessarily applied for data analysis, but by analyzing graphical images the quality of the outcome is appreciated. Indeed, the graphical image may outweigh in importance the parent algorithm. Several, related algorithms will yield graphical images of similar, visual appearance because they are applied to the same data. Each algorithm is analogous to a camera, an instrument for recording a photograph. If taking a picture of a tree for instance, the resultant photograph reveals the tree regardless of the make of the camera. The photograph is carefully studied to infer information about the tree. The camera (algorithm) is no longer important to the analytical process. Emphasizing graphics in the understanding of data is known as **graphical data analysis.**

Data analysis is often considered to be boring in an academic setting. A major purpose of this text is to alter this perception by supplementing data analysis with graphical images to render algorithmic outcomes visually exciting. Accompanied by pictures, data analysis can yield great insight into natural processes. The key is knowing what picture to present when illustrating a concept, process, or outcome. Once again, the objective is to change the perception of data analysis from that of a boring enterprise to one that is interesting, exciting, and meaningful.

The modern age of the Internet, movies on DVD, and computer screens on which graphical images known as icons are clicked to start programs, presents a challenge to educators. Lectures delivered in the classic tradition, reading notes from behind a podium, no longer works to motivate learning. The defense, "It is up to the student to learn," is unfair. The classic style of lecture is so foreign to modern students that its purpose to teach is lost. It is not a visually exciting or meaningful experience.

This is not to say that lectures should be watered down and interlaced with cartoons to be effective. Nothing could be more insulting to the modern student. More to the point, a cultural challenge is necessarily confronted. Anthropologists teach that when reporting on a newly discovered culture, for instance, observations about it must be recorded in the context of that culture, without value judgements from an observer's cultural upbringing. Following this principle of cultural anthropology, the value of visualization in modern culture must be appreciated within the context of this culture, not scorned by those whose upbringing was shaped by different cultural influences.

Visualization is the key to modern education. Meaningful, and substantive illustrations must accompany words, equations, and numbers. This text is an experiment on data visualization.

1.2   Scope

Each chapter of this text consists of four fundamental sections:

- *motivation*—why is the topic of the chapter important to data analysis and visualization;
- *algorithms and analytical tools*—provides mathematical background for the chapter, supplemented by examples of hand calculations that render equations more obvious;
- *demonstration*—a set of Landsat images represents the master data that are analyzed throughout this text in an effort to relate one chapter to the others. Each chapter presents a substantive analysis of these data; occasionally, other data sets are used when appropriate to demonstrate aspects of data analysis that cannot be shown well using these Landsat data;
- *how to repeat analyses and visualizations on your own*—this text is a tool for learning. It represents a challenge to transcend examples that are presented. A computer program, *Visual_Data*, supplied on the CD-ROM at the back of this text, is capable of all analyses and visualizations that are presented. When possible, information is presented showing how to reproduce results and graphics using Microsoft's *Excel* spreadsheet program and Mathwork's MATLAB 5.3 computer analytical environment. A suite of problems at the end of each chapter further challenges students to conduct analyses and construct meaningful images of results.

This text is written in a logical manner, progressing from the fundamental notion of statistical analysis through three-dimensional graphical images of topography. Each chapter builds on previous ones. Topics are to be viewed as interrelated, not disjunct as is often the perception when viewing the order of chapters of some texts. A list of chapter topics illustrates the logical progression of this text.

Univariate data analysis—an important first step. Values of a single variable—copper, silver, gold, electromagnetic reflectance—are inspected for average numerical magnitude (mean), numerical variability with respect to the mean (variance), and how frequently data values of similar numerical size occur in the total set (histogram). The histogram identifies the distribution; once identified, an ideal mathematical model, such as the normal distribution, may be chosen to represent it.

Bivariate data relationships—once an appreciation for single variable values is obtained, the relationship between two variables is investigated. Is gold related to silver in a particular ore deposit? Is conifer species related to soil type? Is the amount of blue light reflected by a rock related to the amount of red light reflected by it? Answers to these questions represent deeper scientific intuition. Methods of regression are shown to be useful for the investigation of these questions.

Multiple variable (multivariate) data relationships—often in the natural sciences data representing more than two variables are collected for analysis; examples include rock geochemistry, in which many elements are measured to infer rock origin and mineralization; water geochemistry, in which many elements are analyzed to infer origin or contamination; and satellite and aircraft imaging using sensors capable of dividing the electromagnetic spectrum into hundreds of slices (bands). Given these multiple variables, how are they collectively intercorrelated? Does this understanding yield insight into natural processes that are not easily interpreted from univariate or bivariate data analysis? These are some of the key issues that are investigated. Analytical techniques, known as principal components methods, are used to assess multivariate data analysis. Although they sound daunting, these methods are applied to yield a graph (a picture or visualization) of these multiple variables that is studied, to infer answers to these questions.

Univariate spatial analysis—transcends the inspection of data value to also examine where the value occurs in space. Both value and location are important to this analytical endeavor. Space may be one, two, three, or multidimensional. Time is necessarily included as a notion of one-dimensional space, and this chapter includes a discussion of time series analysis. More substantive is the introduction to and discussion of spatial autocorrelation. A tool known as the variogram is shown to be useful for visualizing spatial autocorrelation. Another tool, **kriging,** is introduced as a data interpolation method that is based on the variogram.

Multivariate spatial analysis—introduces the notion of spatial cross correlation and the related data interpolation tool known as **cokriging.** When faced with mapping more than one variable in space, cokriging may be the preferred tool to use. But, is cokriging "better" than kriging when mapping multiple variables in space?

Geostatistical simulation—changes the focus from spatial data analysis to spatial data modeling. Random number generation is discussed and shown to be the foundation from which geostatistical simulation builds. Simulation is a game for understanding the impact a sampling scheme has on the ability to visualize the distribution of data value in space. Other games may be played using simulations such as artificial mining to appreciate ore grade fluctuation, or artificial farming to improve prediction of crop yield.

The final two chapters of the book represent capstone summarization of the topics presented thus far. These chapters present advanced visualizations, constructs dependent on many of the analytical tools that are presented throughout the text.

Digital image processing—an excellent forum for visual spatial data analysis. Methods are presented for changing contrast and filtering. Principal components analysis is revisited to show the visual results of multivariate data analysis applied to a set of multispectral, digital images. A demonstration is presented showing how dust devils were identified in images returned to Earth in 1997 by the Mars Pathfinder.

Data compositing—the text ends with visualizations created by combining two or more digital images. Digital image classification is discussed in the context of composites. Principal component images are composited, both for classification and color analysis. The latter offers a visual approach to correlation analysis. Contour line drawing is introduced, then composited with digital image data. And, three-dimensional grid perspectives are introduced as unique ways to visualize a variety of data. Elevation is viewable in three dimensions in the traditional sense. But, estimated values from kriging may also be treated as "elevation" to infer the locations of data highs and lows. Digital satellite data may be draped over elevation data to render a realistic, three-dimensional visualization of terrain.

To reinforce the motivation for this topical presentation, the following questions are forwarded. Their answers are found by progressing in order through the text:

- Is the distribution of a set of data modeled well by a normal distribution? If it is not, can data analysis proceed?
- What form of regression model is appropriate for a set of data?
- What is multivariate data? How is correlation among three or more variables evaluated?
- What insights to multivariate correlation are derived using principal components-type analysis?
- What is spatial correlation? How is this different from bivariate or multivariate correlation? What is kriging?
- When does cokriging yield an improvement over kriging?
- When is geostatistical simulation more appropriately applied than estimation?
- How can digital image processing improve the understanding, moreover visualization of spatial data?
- Can variable and spatial correlations be inferred on a visual basis by compositing two or more digital images, each representing a different variable or information?

1.3   *Nevada_Landsat* Data Set

Throughout this text, the same data set is analyzed to establish a unifying theme. This data set, dubbed *Nevada_Landsat*, is a Landsat TM (thematic mapper) scene, Path 42/Row33, of *Walker Lake, Mineral County, Nevada* that was acquired on July 7, 1984. The Landsat TM scanner acquires images in the following electromagnetic frequencies (known as its **spectral resolution**):

Landsat Thematic Mapper Spectral Resolution							
	Band 1	**Band 2**	**Band 3**	**Band 4**	**Band 5**	**Band 6**	**Band 7**
Name	Visible Blue	Visible Green	Visible Red	Near Infrared (NIR)	Mid-Infrared (Bn5)	Thermal Infrared (Thr)	Mid-Infrared (Bn7)
Wavelength (μm)	0.45–0.52	0.53–0.60	0.63–0.69	0.76–0.90	1.55–1.75	10.4–12.5	2.08–2.35

Band is a term that means "slice," or a discrete segment of the electromagnetic spectrum.

This text is not about remote sensing. Much attention is given, however, to spatial data analysis. Satellite-acquired images are excellent representatives of spatial data, partially explaining the choice of these data. Because Landsat TM images comprise seven unique bands, they are **multispectral,** which also means **multivariate.** Consequently, these images represent interesting sets of data for use in understanding multivariate data analysis, another reason for selecting a Landsat TM image for this text. Landsat images afford an aerial perspective of Earth's surface, aiding the study of geology, botany, hydrology, hydrogeology, geography, archaeology, soil science, and engineering design for urban growth and hazards mitigation. This is yet another reason for choosing these data for analysis.

Although this is not a text on remote sensing, the electromagnetic spectrum is necessarily introduced and explained for a complete understanding of the Landsat TM data. A graph is a convenient tool for illustrating this spectrum. Notice that labels, such as visible, are used to identify particular regions of the spectrum. Moreover, wavelength varies, from smallest, at the left, to largest, at the right (Figure 1.1).

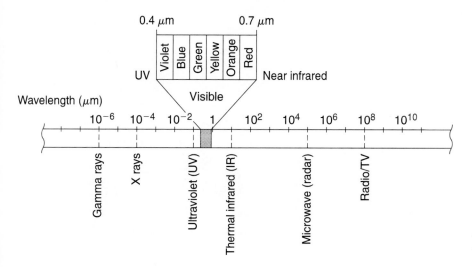

◀ FIGURE 1.1
The electromagnetic spectrum.

Electromagnetic energy (EM) is a two-component (electrical-magnetic) phenomenon that propagates (travels) as a wave. This explains the use of wavelength when discussing the EM spectrum. EM energy, regardless of wavelength, travels at the speed of light. Because it is a wave phenomenon, characteristics of wave refraction, reflection, aliasing, are inherent to EM energy propagation.

Human familiarity with EM energy, once limited, is now extensive. Visible light is one electromagnetic phenomenon that is detectable by the human eye. Other EM phenomena include **ultraviolet** (UV), that causes some minerals to fluoresce and is also responsible for sun burn, **x-rays,** very small wavelength EM energy useful for medical imaging, **thermal infrared** (heat) useful for many human endeavors, moreover warming our planet as Earth converts intense visible EM energy from the sun to heat, **microwave** (radar) used for airborne and space borne imaging, weather imaging, police radar guns, and microwave ovens, and **radio-television** especially useful for communication. There is also the **near** and **mid-infrared** (not the same as heat), useful for the imaging of vegetation and some mineral alteration byproducts, and also used extensively for the design of remote controls for electronic equipment. Clearly, the utilization of the entire electromagnetic spectrum has been an important aspect of human technological development.

Although the human eye is able to detect only visible light, other EM frequencies are visualized by first imaging them, then displaying their image on a computer monitor. The monitor emits visible light with an intensity that is directly proportional to the strength of the imaged EM frequency. This display represents a translation of the EM frequency to the visible frequency. The following seven figures are representative. These are the seven Landsat TM bands of the *Nevada_Landsat* data set.

Notice the differences among these images. Each is a visualization of a different portion of the electromagnetic spectrum. A few observations stimulate inspection. Crops, active vegetation, and north of the lake are dark in the visible bands, but bright in the near infrared band. Healthy, living vegetation is a strong reflector of near infrared energy, whereas most other earth materi-

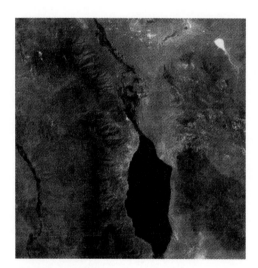

Band 1, Visible Blue

Band 2, Visible Green

Band 3, Visible Red

Band 4, Near Infrared

Band 5, Mid-Infrared

Band 6, Thermal Infrared

Band 7, Mid-Infrared

als are absorbers at this frequency. The playa in the top, right portion of the image is quite bright in most of these bands, indicating strong reflectance over a broad portion of the spectrum. Water has low reflectance in all bands; visible light penetrates water more deeply than longer wavelengths. The shorter the wavelength of EM energy, the deeper is its penetration.

Each of these images is a subsection of the original image. Each display represents 1600 lines of pixels; each line is 1600 pixels in length. A **pixel** is one number (digit) making up a digital image. In the case of Landsat TM images, each pixel's value represents the relative strength of the reflection (emission in the case of Band 6) over a 30m × 30m spatial region (120m × 120m in the case of Band 6). This is the **spatial resolution** of the Landsat TM satellite. Moreover, each pixel is integer-valued, represented by a single byte (8 bits). Consequently, Landsat TM pixels range in value from 0 to 255 (for a total of 256 possible values). This is the **radiometric resolution** of the Landsat TM satellite. Finally, this satellite passes over the same region of Earth every 16 days. This defines the **temporal resolution** of this space borne imaging system. Later on, in Chapter 9, a Landsat TM image of *Walker Lake* acquired on

July 29, 1986 is used to demonstrate composite display of Landsat and digital elevation data. Repeated imaging over time enables the study of change.

With this background information about Landsat TM data in mind, the focus now shifts to the insight each of the following chapters provides into these data.

1.4   Literature

Information about the Landsat satellite is obtained from most texts on remote sensing. Two are used as sources of information for this chapter, Schowengerdt (1997) and Vincent (1997).

Exercises

1. Do you agree with the opening quote to this chapter? Why or why not?
2. Write a summary of your experiences with mathematics since kindergarten. What was your most positive experience with mathematics? What was your most negative experience? Was the negative experience due to a teacher? Was the lack of visualization a factor? Discuss your history with your instructor.
3. Why is the playa in the upper, right portion of the *Nevada_Landsat* images darker in the Band 6 image?

(Hint: Landsat images are acquired at approximately 10:30 a.m. local time. This is a mid summer image. Sunrise occurred at approximately 6:30 a.m. local time).

4. Does UV penetrate water? Describe a laboratory experiment that could be used to verify your answer.
5. Two images, A and B, are shown of *Walker Lake*. One is a visible blue image. The other is a Band 5, mid-infrared image. Which image, A or B, is visible blue?

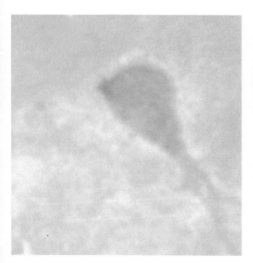

A closer view of the playa is shown. Note its darker tone relative to surrounding terrain.

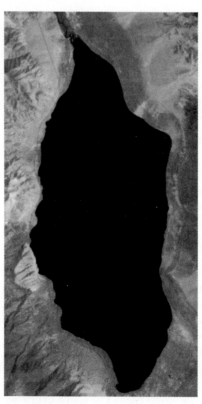

A   B

Univariate Data Analysis

2

Viewing data is the most important starting point in any data analysis process. Inspecting data to understand magnitude (value), variability (differences between data values), and distribution (proportions of the different data values) is crucial for recognizing the adequacy of analytical procedures applied to these data. Looking at only the first several lines in a data file will tell you much about the data. Examples of value, sign (negativeness or positiveness), aspect (integer, real, scientific, character, and so on), variability (as a function of difference in value), and how many values per line of data are readily apparent (Figure 2.1). Glimpsing data in this manner helps to develop a mental picture of whatever the data represents.

135.	65.	92.	84.	156.	177.	88.	1.	1.
132.	65.	92.	81.	150.	175.	85.	1.	7.
133.	65.	93.	81.	150.	175.	86.	1.	13.
131.	64.	91.	79.	151.	174.	86.	1.	19.
138.	67.	95.	79.	154.	174.	89.	1.	25.

▲ FIGURE 2.1
The first five lines of the *Nevada_Landsat_6x.dat* data set. Note that there are nine (9) values per line. The first seven values represent pixel values in Landsat TM bands 1 to 7. The last two values represent the row and pixel number in the original image. These values also represent *Y* and *X* coordinates respectively.

Creating "pictures" from data is the paramount theme of this text. In this chapter, we will explore methods for visualizing univariate data relationships. Univariate analysis examines single variable characteristics. That is, this chapter is devoted to understanding individual variables, moreover the statistics that are used to describe them. Results from more sophisticated analyses have little meaning, if any at all, unless we understand and appreciate univariate parameters and relationships.

2.1 Histograms

One convenient way to visualize univariate relationships, in total, involves computing, then plotting **histograms** (Figure 2.2). This analytical tool graphs data **value** against **frequency,** how frequently data values within a particular numerical interval occur within the data set. For this reason the histogram is sometimes referred to as a **frequency diagram,** although the term histogram is more commonly used. The histogram provides a picture of data **distribution** showing how frequency varies with value. Do data values occur with approximately the same frequency across the range of data values, or do some values occur more frequently than others? This question will become increasingly important as this text proceeds to higher level analytical methods.

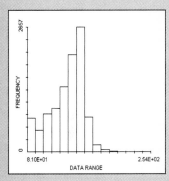

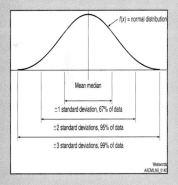

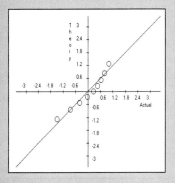

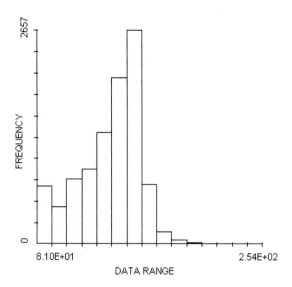

▲ FIGURE 2.2

An example histogram for blue reflectance, *Nevada_Landsat_6x.dat* data set. In this demonstration, 15 bins (intervals) are used to subdivide the data range. This range is 254–81, or 173. The height of each bar represents the frequency (number of data values) for each bin interval.

A histogram can be computed and plotted by hand, but using a computer program is far more convenient. Even so, understanding how the computer program operates is essential to the interpretation of its results. When computing the histogram, the minimum and maximum data values are first identified, then the **range** is computed as the numerical difference between these extremes. The range is then divided into equal-sized **bins,** subintervals of the range. The number of bins is the choice of the data analyst. Consequently, constructing the histogram involves some subjectivity. Using too many bins overly emphasizes data variability (the histogram will appear to be "noisy"). Using too few bins overly compresses data yielding the false sense of low data variability. Try using ten bins to start, and adjust this number depending on the acceptability of the plot. Larger data sets (N in excess of 1000) allow for more bins; smaller data sets will require fewer bins. Once the bin intervals are determined, the number of times (frequency) that values fall into each bin are counted. These counts (frequencies) are plotted against bin value to construct the histogram.

Frequency is an estimate of **probability.** The probability of a data value occurring in a data set is estimated by the frequency for the bin within which it falls, divided by N, the number of values in the data set. Dividing frequency by N yields **relative frequency.** A histogram may be plotted using either frequency or relative frequency. The vertical axis of the plot must be labeled accordingly to enable a correct interpretation of data behavior. Given that frequency and probability are related notions the urge to use the word *identical* in this context is suppressed because probability is a formal mathematical concept defined as the area under the data distribution curve. This curve is known as the **probability density function, $p(x)$,** whereas frequency in the case of the histogram is a discrete approximation of probability. The histogram is a useful tool for estimating the probability for data and for estimating the **probability density function** for the data set.

A histogram is also useful for visually estimating data **mean, variance, and skew**. Mean data value represents the average data numerical magnitude. It can be estimated from the histogram by taking one-half the sum of the minimum and maximum data values (note that this is simply an estimate for the mean and may not precisely equal the actual mean value). The mean is formally calculated as

$$\bar{x} = \frac{1}{N}\sum_{i=1}^{N} x_i$$

[Mean equals the sum of all data divided by the number, N, of values]

The mean is one "picture" of data; it helps us visualize average data magnitude.

But, the mean tells us nothing about how variable the data values are. Two data sets may have the same mean, for example 100. One set, though, may have values ranging from 0 to 200. The other may be associated with values that range between 90 and 110. So, even though both have the same mean, these two sets of data are quite different with respect to **variability.** The statistical measure of variability is known as **variance.** The square root of variance is known as **standard deviation.** A quick estimate of standard deviation is obtained from the histogram as one-half the difference between the minimum and maximum data values.

Variance is formally computed as:

$$S^2 = \frac{\sum_{i=1}^{N}(x_i - \bar{x})^2}{N-1} = \frac{N\sum_{i=1}^{N} x_i^2 - \left(\sum_{i=1}^{N} x_i\right)^2}{N-1}$$

[Variance is equal to the average squared difference between value and mean]

Standard deviation S, is simply the square root of variance. Variance is approximately equal to the average squared difference between data value and mean. This value is not precisely the average squared difference because we divide by $N - 1$ rather than N. Notice that the computation of variance is dependent on the mean. For this reason, the variance is referred to as the second statistical moment with respect to the mean. Variance assesses data variability against the mean. Because the mean is computed using the N data values, one piece of information is theoretically removed from the data set—the mean—leaving one fewer data value on which to base the calculation of variance. Clearly nothing is actually removed from the data set; this discussion is entirely within the realm of theory. But, for correctness in calculation, specifically to avoid bias in analysis, we divide by $N - 1$ when calculating variance to account for the theoretical removal of the mean.

In addition to computing mean, variance, and standard deviation, we may also compute **coefficient of variation.** This coefficient measures variance against the mean as:

$$Coef(Variation) = \frac{S^2}{\bar{x}^2}$$

[Coefficient of variation is equal to variance divided by the square of the mean]

The practice of univariate data analysis is now emphasized. The *Nevada_Landsat* data set consists of seven variables, each representing a different spectral region of the electromagnetic spectrum. This data set is fully described in the Introduction. Univariate analysis is useful for understanding the characteristics of each spectral band. The program *Visual_Data*, is applied to the single data set *Nevada_Landsat_6x.dat*. The Histogram and Statistical Analysis Tool, once activated, queries the user regarding which of the seven variables is to be analyzed. Histogram results for blue reflectance is already presented (Figure 2.2). Results for the remaining 6 spectral bands, visible green, visible red, near infrared, mid-infrared (band 5), thermal, and mid-infrared (band 7) are shown in Figure. 2.3. Moreover, statistical parameters are reviewed in the table on the next page:

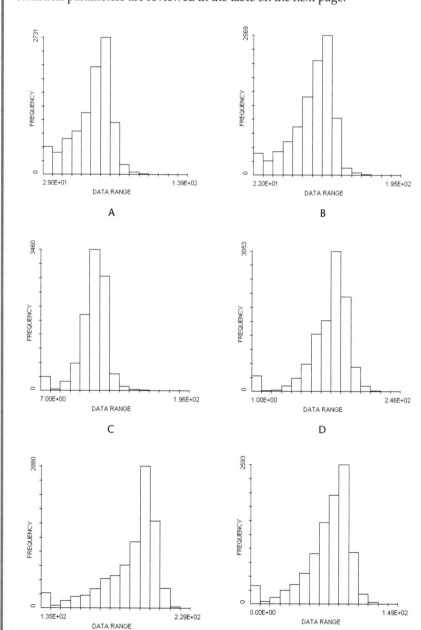

▶ FIGURE 2.3
Histograms for the *Nevada_Landsat_6x* data set: A. visible green; B. visible red; C. near infrared, band 4; D. mid-infrared, band 5; E. thermal infrared; and F. mid-infrared, band 7.

Table 2.1   Statistical summary for the 7 spectral bands associated with the *Nevada_Landsat_6x* data set.

Band	Mean	S^2	S	Coef. Var.	Skew	Coef Skew	Kurt.	Coef Kurt.	χ^2, df = 12	R^2, p-plot n = 10^4	Normal Y/N
Blue	148.5	593	24.4	0.027	−1E+04	0.55	1E+06	3.14	3394	0.976	No
Green	73.5	229	15.1	0.042	−3E+03	0.60	2E+05	3.22	2887	0.978	No
Red	102	641	25.3	0.06	−2E+04	1.15	2E+06	4.03	2572	0.981	No
NIR	86.5	416	20.4	0.056	−1E+04	2.53	1E+06	8.09	1834	0.918	No
Bn5	144	1348	36.7	0.065	−9E+04	3.53	1E+07	7.86	2156	0.986	No
Thr	198	328	18.1	0.008	−9E+03	2.25	5E+05	5.03	4590	0.848	No
Bn7	88	622	24.9	0.08	−2E+04	2.41	2E+06	5.83	2535	0.964	No

In this example, two additional statistical parameters **skew** (from which coefficient of skew is computed) and **kurtosis** (from which coefficient of kurtosis is computed) are shown. Skew is a measure of the symmetry of the data distribution, formally computed as:

$$Skew = N\left[\frac{\sum\limits_{i=1}^{N}(x_i - \bar{x})^3}{(N-1)(N-2)}\right]$$

[Skew is the average cubed difference between value and mean]

Skew is also known as the *third* statistical moment with respect to the mean. Because skew is computed as the cubed deviation between data value and mean, it can be a negative number. Negative values of skew occur when the majority of data fall into bins to the *left* of the mean. Positive values of skew suggest the majority of values are distributed to the *right* of the mean. Each of these conditions represents a data **asymmetry** because data values do not fall equally on either side of the mean data value. Values of skew close to zero, however, suggest more or less equal proportions of data on either side of the mean (data symmetry). By itself, the absolute value of skew has little meaning. Instead, the coefficient of skew should be computed as:

$$Coef(Skew) = \frac{Skew^2}{Variance^3}$$

[The coefficient of skew is equal to the square of skew divided by the cube of variance]

This coefficient measures data symmetry against the variance. Values for the coefficient of skew of 1, or less, are indicative of data symmetry (because the magnitude of skew does not exceed that of variance; squaring skew and cubing variance yield sixth-order magnifications of data value difference with the mean). Values for this coefficient in excess of 1 indicate that skew outweighs in numerical importance the variance, and data asymmetry is consequently pronounced.

Kurtosis is the *fourth* statistical moment with respect to the mean, and is computed as:

$$Kurtosis = N^2 \left[\frac{\sum\limits_{i=1}^{N} (x_i - \bar{x})^4}{(N-1)(N-2)(N-3)} \right]$$

[Kurtosis is equal to the average fourth-order difference
between value and mean]

As with the other statistical parameters, variance is used to assess the magnitude of kurtosis. The coefficient of kurtosis is computed as:

$$Coef(Kurtosis) = \frac{Kurtosis}{Variance^2}$$

[The coefficient of kurtosis is equal to kurtosis divided
by the square of variance]

Kurtosis is a magnification of data variability. The more disperse (different) the data are, the larger is the value of kurtosis (and its coefficient). Another way of visualizing this concept is to think of the width of the histogram. The larger the value of kurtosis, the wider and flatter is the histogram. The smaller the value of kurtosis, the more narrow and taller is the histogram.

2.1.1 Visualizing Univariate Data Through Statistical Parameters: A Summary

Statistical moments represent a small set of numbers that enable us to develop a mental picture of our data. We can estimate or approximate the histogram for our data as a sketch based on values of mean (average magnitude), variance, coefficient of skew, and coefficient of kurtosis. Comparing the values of moments to the actual histogram helps to verify the adequacy of the histogram for representing the distribution of our data. In general:

- **mean** is the average numerical magnitude of all values within a set of data—helps us to picture numerical size;
- **variance** assesses the average difference between value and mean for a set of data—helps us to visualize the total range of data values from smallest to largest and to visualize the uncertainty inherent to our data;
- **(coefficient of) skew** assesses the symmetry of the distribution (histogram) of our data with respect to the mean; a value closer to zero indicates more or less equal percentages of data lesser and greater than the mean; a larger value of the coefficient of skew indicates an asymmetry; negative values of skew indicate the majority of data values are less than the mean and positive values of skew indicate the majority of data values exceed the mean;
- **(coefficient of) kurtosis** is a more severe measure of data variability assessed against the mean value; larger values of this coefficient indicate greater deviation from the mean and consequently a flatter histogram; smaller values of this coefficient indicate lesser deviation from the mean as shown by a taller, more narrow histogram;
- **histogram** is a graph showing how frequently data values occur plotted against data value. An important reason for computing and plot-

ting a histogram is to enable the visualization of the distribution of data values within the set. This distribution is approximated as a smooth curve by connecting the tops of the bars of a bar graph representation of the histogram (Figure 2.2). A distribution curve enables us to define the probability of values within our data set.

Review the preceding analysis of the *Nevada_ Landsat_6x* data set to judge the adequacy of these generalizations.

Are there higher order statistical moments beyond kurtosis, and if so of what value are they for interpreting and visualizing data? Yes, there is no theoretical limit to the order of statistical moments, although the usefulness for data analysis of statistical moments beyond order four is not well established. By reviewing the equations for variance (second-order moment), skew (third-order moment) and kurtosis (fourth-order moment), the equations for fifth-, sixth-, seventh-order moments and so on can be derived. It is difficult to find practical applications for and discussions of statistical moments of higher order than kurtosis. One application for them is discussed in Chapters 5 and 6 on univariate and multivariate spatial analysis. Therein, statistical moments up to tenth-order define the coefficients of Hermite orthogonal polynomials used to transform a data distribution to a normal distribution.

2.2   Data Distributions: Probability Density Functions

The term, distribution, has now been used several times in this chapter. One particular type of distribution, the **normal distribution,** is mentioned in the preceding section. Histograms are graphical pictures of data distributions. Rarely do they depict "perfectly-shaped" distributions. Almost always, varying degrees of asymmetry are apparent. Quite often, "noises" are revealed as bins having unusually high (or low) frequencies, and by data values plotting far away from the mean and the majority of the data. These aberrant values are called **outliers.**

Aberrations, whatever they may be, are a characteristic of real (real-life; actual) data that we must come to accept and accommodate while still making reasonable inferences about the process, or processes, the data represent. Perhaps the most difficult aspect of data analysis is having confidence in one's ability to make reasonable inferences when working with noisy, or otherwise imperfect data.

Mathematical models are functions intended to represent a process adequately enough to yield valid estimates of it. In data analysis, one example of a mathematical model is the **probability density function,** $f(x)$. This model is intended to represent the data distribution, the **shape** of the histogram. Probability, $f(x)$, and relative frequency are interchangeable notions. Probability describes how likely a data value, or event, will occur. If a data value has a frequency of zero, it is not at all likely to occur. Its probability is consequently zero. If every value in a data set is the same, the frequency of that value is N, the size of the data set, and its relative frequency is N/N, or 1. The probability of that value occurring in the data set is also 1. Probability is any value between 0, not likely, and 1, completely likely.

The probability density function, if valid for a set of data, yields a value, $f(x)$, that closely matches the relative frequency for the datum, x. Such functions are "smooth" and filter imperfections from actual distributions. Moreover, probability density functions offer a convenient means for recalling the probability of a datum, x, without needing to access and analyze again the entire data set. This is an example of data **compression,** using a relatively few numbers that at any time may be used to recall the whole.

Of course, accurate data compression and recollection is possible only if a probability density function represents a data distribution well. There are therefore two aspects to consider when modeling a data distribution with a probability density function: 1) shape of the function, where shape also implies a physically meaningful model for the process represented by the data; and 2) how accurately it fits the data distribution. Both of these issues are developed next.

2.2.1   The Normal Distribution

The most widely applied model for data distributions is the normal, or **Gaussian** distribution. Its shape is similar to that of a bell, hence its colloquial name, **bell-shaped curve** (Figure 2.4). This model has properties making it attractive and convenient for data analysis. The normal distribution is symmetrical about the mean. The mean value is the most frequently occurring. Two-thirds of the data values are within one standard deviation of the mean; only one percent of all data values fall outside of three standard deviations from the mean. Skew is identically zero. And the mean, **median** (that data value for which half of the data are less than and half are greater than), and **mode** (the most frequently occurring datum) coincide.

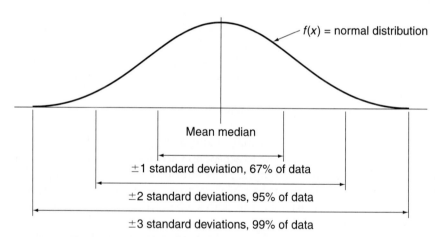

▲ FIGURE 2.4
The normal distribution, also called the Gaussian distribution, or colloquially known as the "bell-shaped" curve.

In equation form, the standard normal distribution is written as:

$$f(z) = \frac{1}{\sqrt{2\pi}}e^{-0.5z^2}; \ for \ which \ z = \frac{x - \bar{x}}{s}$$

[Probability, $f(z)$, is proportional to the exponential function evaluated for one-half the negative of the square of the z-score for x]

The quantity, z, is known as the **z-score** for a datum, x. It has a mean of zero and unit variance. The z-score is a measure of how many standard deviations a datum, x, is away from the mean. Moreover, this score is necessary to the process of determining the goodness of fit of a probability density function to an actual data distribution.

The normal distribution function was first envisioned by Carl Gauss, a mathematician and scientist living in the first half of the nineteenth century. For this reason, the normal distribution is often referred to as the Gaussian distribution. Gauss, a noted astronomer, recognized that when making measurements using a telescope he was rarely perfectly precise. His measures were associated with error. When measuring distances, for instance, Gauss recognized that half the time he overestimated, and the other half he underestimated distance. He also recognized that he was more likely to be closer to the true distance than way off. He envisioned a model for errors that showed a greater proportion of errors symmetrically distributed about a central error of zero (perfect precision), with a greater proportion of errors closer to zero than farther away. What he derived was the foregoing exponential function that is the normal probability density.

2.2.2   Selected Comments About the Normal Distribution

Why is so much attention given to this one type of probability density? There are some ideal properties of the normal distribution that greatly aid data analysis. If data can be shown to be modeled well as a normal distribution, these ideal properties are valid measures. For a normal distribution model:

- the **mean** is that value occurring most frequently; that value having the highest probability of occurring;
- the **mean, median,** and **mode** are all equal in value, in concept all are valid measures of the center of the data distribution. They are therefore considered to be **measures of central tendency;**
- two-thirds of all data values fall within one **standard deviation** of the mean; 95% fall within two standard deviations; and 99% fall within three standard deviations. The z-score, z, for a datum, x, is a measure of how many standard deviations x is away from the mean. Any datum more than two standard deviations from the mean is considered to be a **rare event.** Such a datum has a z-score whose absolute value is 2 or greater;
- the normal distribution's symmetry is perfect with respect to the mean; skew is ideally zero;
- the **expected value,** E, of a datum, x, for a normally distributed set of data is equal to the mean. The expected value is that value, x, for which $f(x)$ is a maximum; is that value, x, in other words that has the highest probability of occurring within the data set; this is written as $E(x)$ = mean (for normally distributed data).

2.2.3   Testing the Normal Probability Density for Goodness of Fit

It is not necessarily valid, if ever, to state that data are normally distributed. More appropriately, we should determine whether the normal distribution function is or is not a good model for our data distribution. If we can show that it is a good model, then we can take advantage of the properties of the normal distribution summarized in the preceding section to make definitive statements about our data. If it is not a good model, we may search for another, more appropriate model, or choose analytical methods that are **nonparametric,** ones that do not depend on some distribution model to represent the data.

A convenient method to use when testing the fit of a distribution model to a data distribution is the **probability plot,** or **p-plot.** Such a plot is constructed by computing the z-scores for the **quantiles** of a data set, then comparing these z-scores to those for the quantiles from a perfectly normal distribution, or some other distribution model that is being tested. The quantiles are found in this way:

- first, **rank-order** the data; that is, list the data from the very smallest datum to the very largest datum;
- once listed in this manner, assign each datum a rank; the rank for the smallest datum is 1, that for the next smallest is 2, and so on. The rank for the largest datum is N, where N is the number of data values in the set.
- the first quantile is that datum having a rank equal to $N/10$;
- the second quantile is that datum having a rank equal to $2N/10$; and so on to
- the tenth quantile, which is the largest datum of rank N.
- In other words, a quantile of a set of data is one representing tenths of the data set; the first quantile is that datum for which 10% of data values are less; the second quantile is that datum for which 20% of the data values are less; and so on. It is of interest to note that the median is the fifth quantile.
- Some texts on statistics refer to quartiles of a data set. These are found in a similar manner by first rank-ordering the data; quartiles represent quarters of the data set; the first quartile is that datum having a rank, $N/4$; the second a rank of $N/2$; the third a rank of $3N/4$; and the fourth quartile is the largest datum having a rank of N. The median and the second quartile are the same.

Once the quantiles are determined for a data set, their z-scores are computed by subtracting from each the mean data value then dividing the difference by the standard deviation. These are plotted against the z-scores for the quantiles of a perfectly normal distribution (Table 2.2). Such a plot is a p-plot, or probability plot. If the normal distribution represents a data set well, z-scores for quantiles of the actual data set plotted against the theoretical values from the normal distribution should plot on, or near the 1:1 line (Figure 2.5 on the next page). Significant deviations from this line show deficiencies in the normal distribution model for the set of data.

Table 2.2   *Z-scores of the standard, normal distribution.*

Quantile	Theoretical z-score
1st	−1.28
2nd	−0.84
3rd	−0.53
4th	−0.25
5th	0.00
6th	0.25
7th	0.53
8th	0.84
9th	1.28
10th	3.50

DEMONSTRATION

Nevada_Landsat_6x Data Set Example

A probability plot for the blue spectral band is plotted in Figure 2.5. Quantitative results for all seven spectral bands are shown in Table 2.1. These quantitative results review the outcome of statistical hypothesis testing applied to the outcome depicted in each p-plot.

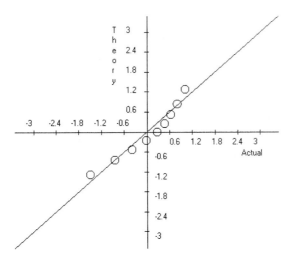

▲ FIGURE 2.5

Probability plot based on histogram analysis of the blue spectral band, *Nevada_Landsat_6x* data set. The plotted line represents a 1:1 correlation between the actual quantiles and those of the normal distribution (theory). This line helps to infer the deviation, if any, from normality.

2.3   Statistical Hypothesis Testing

Beyond visually inspecting the probability plot, how does one say anything substantive about the fit of the normal distribution model to the data distribution? Visual inspection does provide useful insight to the adequacy of the fit. If all pairs, (actual, theoretical), fall on the 1:1 line, the fit is as nearly perfect as is possible. On the other hand, if even one point plots far from this line, where "far" is a subjective assessment, then it is up to the analyst to judge the goodness of fit. This, too, is subjective.

A more objective assessment is possible by applying what is known as **hypothesis testing** to the histogram results to determine goodness of fit of the distribution model. Two such tests are described. One, the **Chi-Square (χ^2) test**, does not depend on the probability plot for analysis. The second test depends solely on results from the probability plot.

In hypothesis testing—some metric—a statistic is computed having a known, well defined distribution function. Distribution functions define the probability of data values, in this case a statistical metric. Data values having low probabilities of occurrence, say five percent probability or less, are deemed to be rare events. The notion of rare events is introduced earlier. Here, this notion forms the basis of the hypothesis test. If the statistical metric is found to be rare, our proposed hypothesis is rejected.

In this chapter, only those hypothesis tests necessary to the current discussion are presented. Other hypothesis tests, such as the **F-test, analysis of variance,** or **ANOVA,** and **t-test** are presented in later chapters when applications are presented enabling a better motivation for these tests.

2.3.1   The Chi-Square Test

Chi-square is a metric for determining the deviation between observed and expected values of data. In application to determining the goodness of fit between a distribution model and actual data distribution, Chi-Square measures the deviation between observed and expected frequencies over all bin intervals of a histogram calculation. In equation form, this is written:

$$\chi^2 = \sum_{i=1}^{K} \frac{(O_i - E_i)^2}{E_i}$$

[The chi-square metric represents the squared difference between an observed frequency, O_i for the i-th bin, and the expected value E_i for this bin, cumulated over K total bins]

In this equation, the expected value E_i, is computed by evaluating the distribution model being tested using the z-score for the center of the i-th bin ($E_i = f(z_i)$). The chi-square metric is consequently useful for testing the fit of any distribution model to a data distribution. The normal distribution model is often the one chosen for testing.

Assuming the normal distribution is the model being tested, the chi-square test proceeds as follows:

Step 1: compute a histogram for the data set using K total bins. Is there a rule for K, the number of bins given the size N of the data set? There is a rule for K, but this rule is not explicitly a function of N. K should be selected such that at least five values occur in each bin. This suggests that K is experimented with to achieve this goal. Some data sets may not allow this goal to be achieved; smaller data sets, or highly skewed data sets may prove difficult for optimizing K. What then? One suggestion is to abandon the use of the chi-square test in favor of the one that is based on results from the probability plot. This alternative hypothesis test is presented momentarily.

Step 2: compute the expected value E_i, of the frequency for the i-th bin as $N(F(z_i) - F(z_{i-1}))$, where z is a z-score, for the ith bin and the bin previous to it, i-1;

Step 3: compute the chi-square statistic using the observed O, and expected E, frequencies;

Step 4: compare the outcome to the table value for K-3 degrees of freedom.

EXAMPLE

Nevada_Landsat_6x Data Set Example

Histogram results for the seven spectral bands are reviewed in Table 2.1. Suppose we wish to test the goodness of fit for the normal distribution applied to the visible blue (Band 1) data distribution. Fifteen (15) bin intervals are used. The following table shows calculations that are necessary to the ultimate calculation of the chi-square statistic:

Actual Z-Scores	Std. Normal Equivalents	Observ	Expected	Chi-Square
-2.774E+00 to -2.691E+00	0.000E+00 to 3.300E-02	5	3.300E+02	3.201E+02
-2.691E+00 to -2.243E+00	3.300E-02 to 5.015E-02	338	1.715E+02	1.615E+02
-2.243E+00 to -1.794E+00	5.015E-02 to 5.871E-02	364	8.555E+01	9.064E+02
-1.794E+00 to -1.345E+00	5.871E-02 to 9.399E-02	466	3.529E+02	3.626E+01
-1.345E+00 to -8.968E-01	9.399E-02 to 1.816E-01	646	8.764E+02	6.055E+01
-8.968E-01 to -4.482E-01	1.816E-01 to 3.238E-01	993	1.422E+03	1.295E+02
-4.482E-01 to 4.581E-04	3.238E-01 to 5.011E-01	1220	1.772E+03	1.720E+02
4.581E-04 to 4.491E-01	5.011E-01 to 6.765E-01	1854	1.754E+03	5.666E+00

```
4.491E-01 to 8.978E-01    6.765E-01 to 8.186E-01    2636    1.421E+03    1.038E+03
8.978E-01 to 1.346E+00    8.186E-01 to 9.061E-01    1142    8.752E+02    8.134E+01
1.346E+00 to 1.795E+00    9.061E-01 to 9.413E-01     237    3.520E+02    3.758E+01
1.795E+00 to 2.244E+00    9.413E-01 to 9.499E-01      56    8.536E+01    1.010E+01
2.244E+00 to 2.692E+00    9.499E-01 to 9.671E-01      22    1.720E+02    1.308E+02
2.692E+00 to 3.141E+00    9.671E-01 to 1.015E+00       8    4.765E+02    4.607E+02
3.141E+00 to 4.332E+00    1.015E+00 to 1.000E+00       5   -1.472E+02   -1.574E+02
Chi Square Statistic = 3.394E+03;  Number of Degrees of Freedom = 12
```

Notes: this table is taken directly from results produced by *Visual_Data*. Therein, Gaussian quadrature, a numerical integration technique, is used to compute the standard normal distribution value for each actual z-score. Some error is noted in this; for instance, notice the calculations for the 15th (last) bin.

Determining the value of the chi-square statistic is the first part of this particular statistical hypothesis test. The next step is to propose the hypothesis. Because the goodness of fit between a model and actual data distribution is the objective, the hypothesis that is tested known as the **null hypothesis** H_0, is stated as follows:

H_0: the data distribution is represented well by the model (e.g., normal) distribution.

Once the null hypothesis is stated, it is also necessary to state the **alternative hypothesis,** H_a. The alternative hypothesis is that which is assumed to be correct should the null hypothesis be rejected. In the case of determining the goodness of fit between a model and data distribution, the alternative hypothesis is simply that the chosen model does not represent the data distribution well.

DEMONSTRATION

Nevada_Landsat_6x Data Set Example

Modeling the visible blue distribution as a normal distribution:

H_0: the visible blue data distribution is represented well by the normal distribution model;

H_a: the visible blue data distribution is not represented well by the normal distribution model.

The chi-square statistic is a **random variable,** an outcome (value) from an experimental process (observation is included), having a finite value and defined probability. This statistic has a distribution known as the **chi-square distribution.** This distribution is not symmetrical; rather, it is skewed to the right (tapers to the right). It is this distribution model that is used to evaluate the null hypothesis against the value of the chi-square statistic. If this statistic's value falls within the rare event portion of the chi-square distribution, then we shall conclude that the null hypothesis must be **rejected** in favor of the alternative hypothesis. If the value of the chi-square statistic does not qualify as a rare event, the null hypothesis is not rejected. A table of chi-square distribution values (Appendix A) is necessary for determining where in the distribution the value of the chi-square statistic falls.

DEMONSTRATION

Nevada_Landsat_6x Data Set Example

Modeling the visible blue distribution as a normal distribution: the value of the chi-square statistic is determined earlier to be 3394. The number of degrees of freedom for this test are $K - 3$. The quantity, 3, is taken away from K, the number of bins, to account for the quantities, N (total data), the mean, and the standard deviation. Referring to Appendix A, the table value for 95% (rare event designation) and 12 degrees of freedom is 23.3. We therefore reject the null hypothesis because the test statistic vastly exceeds this table value. We conclude on this basis that the normal distribution model is not representative of the actual data distribution.

2.3.1.1   Discussion

Given the size, N, of a set of data, the larger N is the more difficult it is to demonstrate the adequacy of the normal distribution model for representing the actual data distribution. Conversely, the smaller N is, the more difficult it is to demonstrate that the normal distribution model is not adequate for representing the data distribution.

In the foregoing calculations applied to the visible blue reflectances from the *Nevada_Landsat_6x* data set, N is quite large—10,000. Recall the second step in the chi-square hypothesis test. The expected value, E, is directly proportional to N. The larger N is, the larger the expected value will be. Even the slightest deviation from true normal distribution behavior on the part of the actual data is magnified by larger N in the chi-square test. On the other hand, if N is small, slight deviations from normality exert lesser impact on the analysis.

DEMONSTRATION

Given the following 20 random numbers:

$$.9\ .3\ .6\ .6\ .8\ .6\ .2\ .1\ .9\ .2\ .4\ .2\ .1\ .3\ .6\ .5\ .9\ .3\ .8\ .7$$

Let's use $K = 4$ bins. The outcome is:

Actual Z-Scores	Std. Normal Equivalent	Observ	Expected	Chi-Square
-1.443E+00 to -1.082E+00	0.000E+00 to 1.383E-01	5	2.767E+00	1.803E+00
-1.082E+00 to 0.000E+00	1.383E-01 to 5.009E-01	5	7.251E+00	6.987E-01
0.000E+00 to 1.082E+00	5.009E-01 to 8.617E-01	5	7.216E+00	6.804E-01
1.082E+00 to 1.443E+00	8.617E-01 to 1.000E+00	5	2.767E+00	1.803E+00

Chi Square Statistic = 4.984E+00;   Number of Degrees of Freedom = 1

In this particular example, the data obey a *uniform distribution*, one in which the frequency for any one data value is equal to the rest, yet the Chi-square test does not result in a rejection of the null hypothesis. For 1 degree of freedom, at 0.05%, the table value is 5.024, larger than the test outcome of 4.984. The smaller the data set, the less likely the null hypothesis will be rejected in the chi-square test. The hypothesis test based on results from the probability plot yield a squared correlation coefficient of 0.77 for these data. For 20 degrees of freedom, this coefficient should be larger than 0.86 for the normal distribution model to be considered representative. Results from the probability plot indicate that the normal distribution model is not representative of these data (as should the outcome be because these data are better modeled by a uniform distribution). The hypothesis test based on results from the probability plot is described in the next section.

2.3.2   Hypothesis Test for the Probability Plot

As an alternative to the chi-square test, a more direct test is possible using the theoretical and actual z-scores of the data quantiles to compute a **correlation coefficient, r**. A definition and detailed explanation of correlation (between two variables) is deferred to the next chapter. For the present discussion, the following equation for r is proposed:

$$r = \frac{N \sum_{i=1}^{N} (O_i E_i) - \sum_{i=1}^{N} O_i \sum_{i=1}^{N} E_i}{S_O S_E}$$

[Correlation coefficient, r, for this application, is a ratio; the numerator represents the cumulative sum of the products of the theoretical and actual z-scores, less the product of the two sums, actual and theoretical; the denominator is the product of the standard deviations of the actual and theoretical z-scores]

Notation in this formula is as follows: O and E represent respectively the actual and theoretical z-scores from the probability plot. S represents the standard deviation, with subscripts, O and E, designating actual and theoretical z-scores respectively. Once the correlation coefficient, r, is computed, its square, r^2, is determined and compared to table values (Appendix B) to judge the adequacy of the null hypothesis. For this hypothesis test, the null and alternative hypotheses are those that were stated for the chi-square test.

DEMONSTRATION

Nevada_Landsat_6x Data Set Example

The squared correlation coefficient is 0.976. The size, N, of the data set is 10000 (for *Nevada_Landsat_6x*). Using Appendix B, for such a large data set, this squared correlation coefficient should be larger than 0.991. We therefore reject the null hypothesis in this case, and conclude that the normal distribution model is not representative.

2.3.3   So, Which Hypothesis Test Should be Used?

Confusion is often the result for students when textbooks attempt to cover many alternative methods of analysis. The chi-square test is covered here because many instructors and practitioners know it and use it. But the chi-square test will yield different results depending on how many bins are selected when computing the histogram. This test also requires a minimum frequency of five for each bin. This is an optimal objective commensurate with the use of the chi-square statistic and its conformity to the chi-square distribution. The chi-square distribution model is an optimal fit if a minimum frequency of five values per bin is achieved.

Hypothesis testing based on the probability plot does not depend on the histogram. Instead, the data are rank ordered. The ten data quantiles are then determined, converted to z-scores, and plotted against z-scores for quantiles of the normal distribution. Quantiles do not change when the number of bins used to compute a histogram is changed. No minimum frequency is required when optimizing the test. For these reasons, an analyst may prefer the hypothesis test based on results from the probability plot when evaluating the goodness of fit of the normal distribution model.

Often, the chi-square test and the test based on the probability plot lead to the same conclusion, acceptance or rejection of the null hypothesis. Occasionally, these tests yield conflicting results, one suggesting the null hypothesis should be rejected, the other showing no evidence to reject the null hypothesis. In this case, the goodness of fit of the normal distribution model to the data distribution is most likely marginal. The chi-square test is more likely to suggest rejection of the null hypothesis if the data set size, N, is large. Consequently, for larger N data sets, results from the probability plot can be given more weight. For smaller N data sets, deciding on which test results to believe is up to the data analyst.

2.4   Other Distribution Functions and Suggested Applications

2.4.1   The Poisson Distribution

Suppose we are interested in predicting:

- the number of rock joints per meter in a particular region
- the number of tornados per day in Lubbock, Texas
- the number of dust devils per day in a region of Mars

These three examples represent the number of something per interval. Such data are modeled very well using a **Poisson** distribution. This distribution function is also a function of the exponential:

$$p(x) = \frac{e^{-\mu}\mu^x}{x!}$$

[probability of x, a whole number quantity, is equal to the product of the exponential evaluated for the negative of the average number of events to have been observed and this average raised to the x power; this product divided by x-factorial]

Suppose an average, μ, of 0.5 rock joints per meter are observed at Yosemite National Park. What is the probability that along any given path, 2 joints, x, will be crossed in one stride (1 meter)? In this example, $p(2) = (e^{-0.5})(0.5^2)/(2 \times 1) = 7.6\%$; that is, approximately 8% of our strides will cross 2 joints.

> **DEMONSTRATION**
>
> **Mars Dust Devils**
>
> Mars Pathfinder imaged five dust devils over a two day period (Metzger, et. al., 1999). Suppose this adequately characterizes dust devil frequency on Mars (2.5 dust devils per day). What is the probability that twelve dust devils will occur in any given day? $P(12) = (\text{EXP}(-2.5) \times 2.5^{12})/12! = 1.02\text{E}{-}05$, a seemingly small probability. Examine the Mars Global Surveyor images from 1999 in Figure 2.6 on the next page.

2.4.2   The Exponential Distribution

Discussion has been presented about the symmetry of data distributions. The normal distribution is symmetrical with a value of skew that is ideally zero. Many data distributions though, are not symmetrical and display

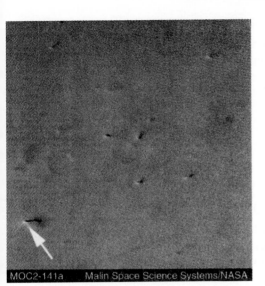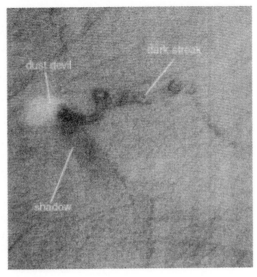

▲ FIGURE 2.6

Two images from Mars Global Surveyor (courtesy Malin Space Science Systems and NASA Jet Propulsion Laboratory (JPL)). The image on the left, MOC2-141a, is from July, 1999. The image on the right was acquired later in 1999 and shows a single dust devil and the dark streak that it has formed along the surface as it entrains lighter colored dust consequently exposing darker material underneath.

There are at least 12 dust devils in the left-hand image (MOC2-141a, courtesy NASA/JPL and Malin Space Science Systems). The fact that this many dust devils were imaged at one time suggests the probability of 12 dust devils equaling approximately 1E-05 is too low. This further suggests the average number of dust devils per day on Mars is much higher than 2.5. The right-hand image shows a single dust devil (MOC2-220) that has formed a dark streak. This was the first direct evidence that dust devils leave this kind of signature across terrain. A recent press release from NASA JPL stated that 5 to 10 dust devils are commonly imaged at one time by Mars Global Surveyor. This may suggest that the average number of dust devils for any one location at any given time is 7.5. Using this value, then p(12) for the left-hand image, assuming a Poisson process, is equal to 0.036 (3.6%). The probability, p(1), for the right-hand image showing a single dust devil is 0.004 (0.4%). And, the probability, p(5), for an image containing 5 dust devils is 11%. These values seem reasonable. Twelve dust devils at one time certainly is possible, and a probability of 3.6% suggests this possibility, but this also suggests that this number is relatively rare. Imaging only a single dust devil is less likely than imaging 12. Five dust devils at one time is more common than 12. All of these inferences from the Poisson model match what Mars Global Surveyor has imaged. A mean value of 7.5 therefore seems to be a reasonable estimate for average dust devil frequency for any given locale at any given time on Mars. Dust devils may be the dominant erosion and transport mechanism on Mars given its current atmospheric conditions and dynamics.

skew. Such an asymmetrical distribution is often suggestive of exponential decay. The **exponential** distribution may be a useful model:

$$f(x) = \frac{e^{-\frac{x}{\bar{x}}}}{\bar{x}}$$

[probability of x is equal to the exponential evaluated for the negative of the ratio, x over the mean of x; this outcome divided by the mean of x]

Notice that this distribution model depends only on knowledge of the mean value of x, and the quantity, x. No measure of variability is included. For example, suppose x is represented well by an exponential distribution. Further, suppose the mean value of x is 5. What is the probability that x is 3? In this case, $f(3) = (e^{-0.6})/5 = 0.11$, or 11%.

2.4.3 The Log-normal Distribution

Often, data are stated or assumed to be **log-normal,** or **log-normally distributed.** Such data are not represented well by the normal distribution model. Yet, if the data are transformed to their logarithms (e.g., natural logarithms), then the normal distribution model represents these logarithms well. No special equation is necessary for the log-normal distribution. The equation is that for the normal distribution, except data are transformed to log values first.

2.5 How Do I Reproduce Results in this Chapter, or Analyze My Own Data. . .

2.5.1 . . . Using *Visual_Data?*

Visual_Data is the program supplied with this text on the CD-ROM included at the back. Once the program is installed, you may use it by clicking START on the Windows 95/98/Wp/2000/NT main window, choosing Programs, then moving your mouse down to *Visual_Data*, then clicking the *Visual_Data* label that appears.

Visual_Data's start window (called a Splash Form) gives you three options: continue with a brief tutorial, continue without the tutorial, or exit the program. It is suggested that you use the "continue with brief tutorial" option until you are used to the program. This tutorial simply explains how to get started using the program.

Once you continue beyond the splash form, the main program window appears. The upper box is a Rich Text Box. It allows full editing capability as well as data file creation. To open an existing data file, such as the nv6x.dat data set that is analyzed in this chapter, with the CD at the end of this text loaded in your CD-ROM drive (you may also use your DVD drive if your computer is so equipped), click File on the main menu, then choose Open. Then, using your mouse, change the drive letter to match your CD-ROM drive letter. Double click the folder, (\data\nvlandat\), then double click nv6x.dat. The data should now appear in the upper box. There are nine values on each line of this data file. The first seven values represent pixel values (reflectances) in the seven Landsat TM spectral bands. The last two values represent *y* and *x* coordinates, respectively.

To obtain a histogram of any one of the first seven values, click Tools on the main menu, then click Histogram Analysis and Statistics, then respond to the following questions interactively:

1. Have you opened a data file? *Yes*
2. Log transform option *No,* for nv6x.dat; but, if you are curious whether your data are log normal, then respond yes.
3. How many data values appear on each line of your data file? 9
4. Which one do you want a histogram for? 1, 2, 3, 4, 5, 6, or 7; example: 1 = visible blue.
5. How many bins? Your choice; 15 are used for examples presented in this chapter.

Once you respond to the number of bins, a progress bar appears telling you the progress of the calculations. This progress bar will cycle twice—once when computing the actual histogram, and once when the data are rank ordered to determine quantiles for the probability plot. The second cycle takes longer than the first.

Once the calculations are finished, two plots are generated: the histogram and the probability plot. You may print each, or return to the main program. When returning to the main program, results from the histogram calculation are printed in the lower box of the main window. You may scroll through this box, or save these results to a file. To do this, click File, then choose Save As.

2.5.2   . . . Using *Microsoft Excel*?

Excel is a popular spread sheet program that has graphing and data analysis tools. First, verify that your data analysis tools are activated in *Excel*. Click Tools on the main menu. Does the item, **Data Analysis** appear in the list of tools? If not, while still in the Tools list, click Add Ins and add check marks in front of **Analysis Toolpak** and **Analysis Toolpak - VBA,** then click ok. Verify that Data Analysis appears in the Tools list.

As a first step after *Excel* is started, load the data set, nv6x.dat, by clicking File on the main menu, then Open, set the drive letter to your CD-ROM drive (or DVD drive depending on your hardware configuration), then double click the folder, \data\nvlandat\, set file type to All Files, then double click nv6x.dat. At this point, *Excel* shows a portion of this data file and indicates that it is fixed width. This is normal. Simply click Finish and the data should load as columns into *Excel*, with visible blue assigned to column A, visible green to column B, and so on.

To obtain a histogram of visible blue reflectance, for instance, click Tools, then Data Analysis, then Histogram, then Ok. The essential item that must be described in the Histogram window is Input Range. Simply click the letter, A, that heads the column for visible blue. The symbol, $A:$A, should appear in the Input Range window. Now, put a check in the small box in front of Chart Output (the graph of the histogram), then click Ok and the histogram should appear as shown below:

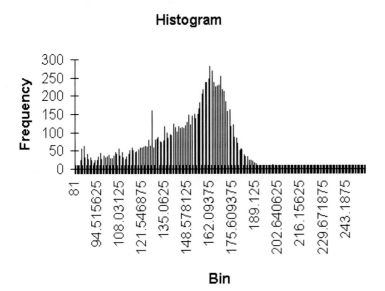

Click on the histogram plot to activate it (small square symbols appear at the edges of the plot); move your mouse to any one of the small squares to resize the plot.

Statistical parameters of the visible blue reflectances are obtained in *Excel* by clicking Tools on the main menu, then Data Analysis, then Descriptive Statistics. Once again, the Input Range must be specified. Simply mouse click the letter, A, heading the column for visible blue, and descriptive statistics as shown below are reproduced:

Mean	148.5273
Standard Error	0.243465753
Median	155
Mode	#NUM!
Standard Deviation	24.34657533
Sample Variance	592.7557303
Kurtosis	0.134866005
Skewness	−0.740548013
Range	173
Minimum	81
Maximum	254
Sum	1485273
Count	10000

2.5.3   . . . Using MATLAB Student Version 5.3?

Many instructors and students prefer the package, MATLAB, for data analysis and graphing. To obtain a histogram of the *Nevada_Landsat_6x.dat* data set using MATLAB, start MATLAB, then use the following commands:

```
>>cd [CD ROM drive letter]:        (Example: >>cd E:)
>>cd Nevada_Landsat_Data
>>load Nevada_Landsat_6x.dat
>>X=Nevada_Landsat_6x.dat
```

To obtain a histogram of the visible blue reflectances, then

```
>>hist(X(:,1), 15)
```

yields a histogram of the first column of the *Nevada_Landsat_6x.dat* data set using 15 bins:

Descriptive statistics for the visible blue reflectances are obtained using MATLAB as follows:

```
(mean):                           >>mean(X(:,1))    (ans=148.5273)
(variance):                       >>var(X(:,1))     (ans=592.7557)
(median):                         >>median(X(:,1))  (ans=155)
(coefficient of skew):            >>skewness(X(:,1)) (ans=-0.7404)
(standard deviation):             >>std(X(:,1))     (ans=24.3466)
(coefficient of kurtosis): >>kurtosis(X(:,1)) (ans=3.1342)
```

Answers (ans) are for the visible blue reflectances of the *Nevada_Landsat_6x.dat* data set. Commands after the symbol, >>, are those required by MATLAB for correct execution.

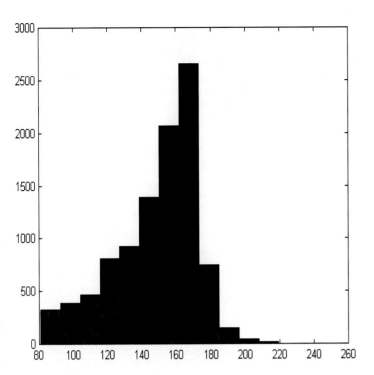

2.6   Literature

A complete bibliography is presented at the end of this text. In writing this chapter, the following references were inspirational: Berthouex and Brown (1994); Davis (1986); Jensen, et. al. (1997); Kreyszig, 6th ed. (1988); McCuen (1985); Rao and Hamed (2000); and von Mises (1957). Additionally, *Middleton (2000)* is an excellent reference on data analysis using MATLAB and was the primary resource for the MATLAB examples presented in this chapter. The MATLAB demonstration was produced using MATLAB 5.3, Student Version, Instructor Evaluation Copy. *Microsoft Excel* examples were produced using *Microsoft Office 97*. *Microsoft Visual Basic 6.0* was used to develop the program, *Visual_Data*. Principal references that were relied upon for *Visual_Basic* are Halvorson (1998), Brierley, et. al. (1998), and Kerman and Brown (2000).

Exercises

1. Show that the following two equations for computing variance are equivalent:

$$S^2 = \frac{1}{N-1}\sum_{i=1}^{N}(x_i - \bar{x})^2; S^2 = \frac{1}{N-1}\left(\sum_{i=1}^{N}x_i^2 - \frac{1}{N}\left(\sum_{i=1}^{N}x_i\right)^2\right)$$

2. Compute: mean, variance, standard deviation, coefficient of variation, skew, coefficient of skew, kurtosis, and coefficient of kurtosis for the following data sets:

   a. 1   2   3   4   5
   b. 10.9 20.2 30.1 25.9 14.8

   c. −7   −8   −9   −10   −11
   d. 102   225   336   91   408
   e. 0.1   0.2   0.3   0.4   0.8
   f. 0.001   0.004   0.006   0.003   0.002
   g. 10,564 22,591 33,456 17,654 8,571
   h. −0.02   0.01   0.03   −0.05   0.03
   i. 2.2   3.3   3.3   3.3   4.4
   j. 1.1   1.1   1.1   2.2   3.3

3. Using the program, *Visual_Data* (or *MATLAB*, or *MS Excel*), repeat the analysis of the *Nevada_Landsat_6x.dat*

data set that is written to the CD-ROM in the directory, \Nevada_Landsat_Data\ (\data\nvlandat\).

4. Using the program, *Visual_Data*, extract a data set from one of the digital images in the Image_Library folder that is supplied on your CD-ROM. Extraction is accomplished by starting *Visual_Data*, then clicking Tools, then moving the mouse pointer down to Digital Image Analysis, then clicking the Image Analysis, then click Extract Pixels ... option. Then, specify the row and pixel region that you want to extract. Please keep your extraction fairly small. For example, extracting a small, 50 row × 50 pixel region yields a data set of 50 × 50, or 2500 values. It is suggested that your extractions be no larger than 100 × 100. Once your extraction is finished, return to the main program and save your data to a file by clicking File, then choosing Save (not Save As). Once your file is saved, click Tools, then click Histogram analysis and statistics and obtain the histogram and statistical summary for this image extraction. How does this analysis compare to the one for the *Nevada_Landsat_6x.dat* data set?

5. Analyze the data extracted in 2.4 using MATLAB or *MS Excel*.

6. Compute a histogram for the following data sets. For each, use either the Chi-Square test, or the hypothesis test based on the probability plot to determine if a normal distribution model is representative. Complete this problem by hand.

   a. Data are taken from DeWijs (1951); percent zinc in a sphalerite-quartz vein:

17.7	17.8	9.5	5.2	4.1	19.2	12.4	15.8	20.8	24.1
14.7	21.6	12.8	11.9	35.4	12.3	14.9	19.6	10.6	15.1
15.6	9.3	8.1	13.5	30.2	29.1	7.4	12.3	13.6	9.5
13.1	27.4	8.8	11.4	6.4	11.0	11.4	14.1	20.9	10.6
15.3	24.0	12.3	7.8	9.9	20.7	25.0	19.1	13.1	27.4
15.2	12.2	10.1	12.3	16.7	18.6	6.0	10.6	11.3	4.7
10.9	6.0	7.2	5.6	8.9	5.8	8.9	6.7	7.2	9.7
10.8	17.9	10.9	13.7	22.3	10.2	5.1	13.9	9.0	10.6
13.8	6.5	6.5	10.6	10.6	23.0	21.8	32.8	30.2	30.8
33.7	26.5	39.3	24.5	24.9	23.2	16.0	20.9	10.3	22.6
16.2	22.9	36.9	23.5	18.5	16.4	17.9	18.5	13.6	7.9
31.9	14.1	7.1	3.9	3.7	22.5	27.6	17.3		

   b. Data are taken from Carr (1995); average U.S. temperature (Celsius), 1895–1987:

10.7	11.55	11.20	11.18	11.00	11.95	11.45	11.30	10.70	11.00
10.9	11.39	11.2	11.55	11.25	11.75	11.50	10.4	11.2	11.4
11.2	10.8	10.35	11.45	11.30	10.95	12.40	11.50	11.39	10.70
11.85	11.50	11.60	11.50	10.90	11.55	12.25	11.30	11.95	12.60
11.4	11.55	11.25	12.00	12.05	11.35	11.80	11.35	11.50	11.40
11.30	11.90	11.40	11.20	11.50	11.15	11.00	11.70	12.15	12.15
11.30	11.60	11.50	11.40	11.50	11.10	11.38	11.38	11.50	11.15
11.15	11.05	11.20	11.00	11.10	11.15	11.15	11.00	11.50	11.50
11.05	11.05	11.65	10.80	10.70	11.50	11.95	11.00	11.30	11.38
10.90	12.10	12.10							

   c. Average July high temperature, New York City, New York, 1876–1992:

79	79	79	77	78	81	78	76	78	78
78	77	76	76	78	80	79	78	79	81
80	79	81	79	82	80	77	76	78	80
81	80	80	77	79	81	79	79	79	79
80	79	79	78	81	82	79	80	80	80
79	78	79	81	82	83	83	81	80	80
82	81	81	82	80	82	81	82	85	80
80	81	82	82	78	82	83	83	81	83
78	82	80	83	80	84	79	79	81	80
83	80	83	81	83	82	82	83	82	79
80	82	79	82	85	81	81	85	79	81
80	81	84	80	81	83	79			

   d. Carbon dioxide (ppmv) as measured at Mauna Loa Observatory, Hawaii (see next page).

   e. Concrete tensile strength, 100 samples, from Kreyszig (1988):

320	350	370	320	400	420	390	360	370	340	380	340
390	350	360	400	330	390	400	360	340	350	390	360
350	350	360	350	360	390	410	360	440	340	390	370
380	370	350	400	380	370	330	340	400	330	350	370
380	370	340	350	390	350	350	320	330	350	380	410
360	380	330	350	360	390	360	390	360	360	350	370
360	390	340	380	300	370	340	400	320	300	400	380
370	400	360	370	330	340	370	420	370	340	420	370
360	340	370	360								

7. Create a file for each data set listed in problem 2.6. Then, verify your hand calculations using *Visual_Data*, MATLAB, or *MS Excel*. If there is a significant skew in any of these data sets, given what the data are, what is the scientific significance of this skew? If the normal distribution model is representative of the data, given what the data are, what is the scientific significance of this realization? Hint: examine the directory on the CD, \data\othrdata\.

Year	Jan.	Feb.	March	April	May	June	July	Aug.	Sept.	Oct.	Nov.	Dec.	Annual
1958	-99.99	-99.99	315.56	317.29	317.34	-99.99	315.69	314.78	313.05	-99.99	313.18	314.50	-99.99
1959	315.42	316.31	316.50	317.56	318.13	318.00	316.39	314.65	313.68	313.18	314.66	315.43	315.83
1960	316.27	316.81	317.42	318.87	319.87	319.43	318.01	315.74	314.00	313.68	314.84	316.03	316.75
1961	316.73	317.54	318.38	319.31	320.42	319.61	318.42	316.63	314.83	315.16	315.94	316.85	317.49
1962	317.78	318.40	319.53	320.42	320.85	320.45	319.45	317.25	316.11	315.27	316.53	317.53	318.30
1963	318.58	318.92	319.70	321.22	322.08	321.31	319.58	317.61	316.05	315.83	316.91	318.20	318.83
1964	319.41	-99.99	-99.99	-99.99	322.06	321.73	320.27	318.53	316.54	316.72	317.53	318.55	-99.99
1965	319.27	320.28	320.73	321.97	322.00	321.71	321.05	318.71	317.66	317.14	318.70	319.25	319.87
1966	320.46	321.43	322.23	323.54	323.91	323.59	322.24	320.20	318.48	317.94	319.63	320.87	321.21
1967	322.17	322.34	322.88	324.25	324.83	323.93	322.38	320.76	319.10	319.24	320.56	321.80	322.02
1968	322.40	322.99	323.73	324.86	325.40	325.20	323.98	321.95	320.18	320.09	321.16	322.74	322.89
1969	323.83	324.26	325.47	326.50	327.21	326.54	325.72	323.50	322.22	321.62	322.69	323.95	324.46
1970	324.89	325.82	326.77	327.97	327.91	327.50	326.18	324.53	322.93	322.90	323.85	324.96	325.52
1971	326.01	326.51	327.01	327.62	328.76	328.40	327.20	325.27	323.20	323.40	324.63	325.85	326.16
1972	326.60	327.47	327.58	329.56	329.90	328.92	327.88	326.16	324.68	325.04	326.34	327.39	327.29
1973	328.37	329.40	330.14	331.33	332.31	331.90	330.70	329.15	327.35	327.02	327.99	328.48	329.51
1974	329.18	330.55	331.32	332.48	332.92	332.08	331.01	329.23	327.27	327.21	328.29	329.41	330.08
1975	330.23	331.25	331.87	333.14	333.80	333.43	331.73	329.90	328.40	328.17	329.32	330.59	330.99
1976	331.58	332.39	333.33	334.41	334.71	334.17	332.89	330.77	329.14	328.78	330.14	331.52	331.98
1977	332.75	333.24	334.53	335.90	336.57	336.10	334.76	332.59	331.42	330.98	332.24	333.68	333.73
1978	334.80	335.22	336.47	337.59	337.84	337.72	336.37	334.51	332.60	332.38	333.75	334.78	335.34
1979	336.05	336.59	337.79	338.71	339.30	339.12	337.56	335.92	333.75	333.70	335.12	336.56	336.68
1980	337.84	338.19	339.91	340.60	341.29	341.00	339.39	337.43	335.72	335.84	336.93	338.04	338.52
1981	339.06	340.30	341.21	342.33	342.74	342.08	340.32	338.26	336.52	336.68	338.19	339.44	339.76
1982	340.57	341.44	342.53	343.39	343.96	343.18	341.88	339.65	337.81	337.69	339.09	340.32	340.96
1983	341.20	342.35	342.93	344.77	345.58	345.14	343.81	342.21	339.69	339.82	340.98	342.82	342.61
1984	343.52	344.33	345.11	346.88	347.25	346.62	345.22	343.11	340.90	341.18	342.80	344.04	344.25
1985	344.79	345.82	347.25	348.17	348.74	348.07	346.38	344.51	342.92	342.62	344.06	345.38	345.73
1986	346.11	346.78	347.68	349.37	350.03	349.37	347.76	345.73	344.68	343.99	345.48	346.72	346.97
1987	347.84	348.29	349.23	350.80	351.66	351.07	349.33	347.92	346.27	346.18	347.64	348.78	348.75
1988	350.25	351.54	352.05	353.41	354.04	353.62	352.22	350.27	348.55	348.72	349.91	351.18	351.31
1989	352.60	352.92	353.53	355.26	355.52	354.97	353.75	351.52	349.64	349.83	351.14	352.37	352.75
1990	353.50	354.55	355.23	356.04	357.00	356.07	354.67	352.76	350.82	351.04	352.69	354.07	354.04
1991	354.59	355.63	357.03	358.48	359.22	358.12	356.06	353.92	352.05	352.11	353.64	354.89	355.48
1992	355.88	356.63	357.72	359.07	359.58	359.17	356.94	354.92	352.94	353.23	354.09	355.33	356.29
1993	356.63	357.10	358.32	359.41	360.23	359.55	357.53	355.48	353.67	353.95	355.30	356.78	356.99
1994	358.34	358.89	359.95	361.25	361.67	360.94	359.55	357.49	355.84	356.00	357.59	359.05	358.88
1995	359.98	361.03	361.66	363.48	363.82	363.30	361.94	359.50	358.11	357.80	359.61	360.74	360.91
1996	362.09	363.29	364.06	364.76	365.45	365.01	363.70	361.54	359.51	359.65	360.80	362.38	362.69
1997	363.23	364.06	364.61	366.40	366.84	365.68	364.52	362.57	360.24	360.83	362.49	364.34	363.82

Atmospheric concentrations of CO2 are expressed in parts per million (ppm) and reported in the preliminary 1997 SIO manometric mole fraction scale. Missing values are denoted by −99.99. In years where one monthly value is missing annual values were calculated by substituting a fit value (4-harmonics with gain factor and spline) for that month and then averaging the twelve monthly values. Source: C.D. Keeling and T.P. Whorf, Scripps Institution of Oceanography (SIO), University of California, La Jolla, California USA 92093-0220, August 1998.

3

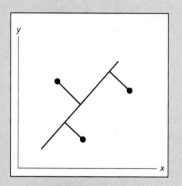

Bivariate Data Analysis

Now that we have examined the nature and characteristics of a single variable, relationships between two variables are visualized. Studying intervariable relationships between two variables is known as *bivariate* data analysis. Although briefly introduced in Chapter 2, bivariate data analysis is the foundation for defining the correlation between two variables. If the correlation between two variables is "good," we may be able to accurately predict the value of one variable given the value of the other. Or, by examining correlation, we may discover a causal relationship between two processes that was not well understood. In the next chapter, correlation is the fundamental basis for examining relationships among three or more variables.

Because creating "pictures" from data is the paramount theme of this text, graphical methods for visualizing *bivariate* data relationships, with a particular emphasis on correlation, are explored in this chapter.

3.1 Correlation

3.1.1 The Method of Least Squares

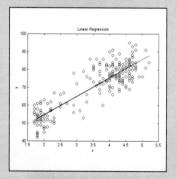

Karl Gauss was a remarkable individual who lived in the early nineteenth century. When only ten years old, his teacher gave the class an assignment to sum the whole numbers between 1 and 100. Students only had small, black slates on which to write using small pieces of white chalk. Most started scratching away, adding $1 + 2$ to get 3, another 3 to get 6; another 4 to get 10; and so forth. Gauss, in contrast, wrote but a single number on his slate and turned it in, the first to do so and only a few seconds after the teacher gave the assignment.

Eyeing Gauss with considerable suspicion, the teacher waited for the rest of the class to finish before grading the slates. All answers were incorrect—except for one—with wrong answers earning the respondent a whack from the teacher's cane. The one right answer, 5050, was on the bottom slate, that which was turned in by Karl Gauss.

"How were you able to arrive at this answer so quickly?" the teacher asked in amazement. Gauss replied that he could visualize patterns in numerical sequences. For instance, he explained that if the assigned problem is written in two ways, forward and backward, then if each pair of numbers across the two sequences is summed as follows:

$$
\begin{array}{ccccccccc}
1 & + & 2 & + & 3 & + & \cdots & + & 100 \\
\underline{100} & + & \underline{99} & + & \underline{98} & + & \cdots & + & \underline{1} \\
101 & + & 101 & + & 101 & + & \cdots & + & 101 = 10100
\end{array}
$$

then, 100 total pairs are obtained, each summing to 101 for a total amount of 10,100. But, because the original problem was written twice, this amount is exactly twice the sum of the whole numbers, 1 to 100. So, the actual sum is (10,100)/2, or 5050.

Gauss contributed much to the science of numbers (mathematics), and the sciences in general. One contribution, that of the normal distribution, was reviewed in the previous chapter. That contribution was one important outcome of his **theory of errors.** When making observations through his telescope, and attempting angular measures between stars or planets, he realized that his measures were associated with error. Recognizing that he was more likely to be closer to the true value, and less likely to be farther away, led Gauss to the formal derivation of the normal distribution as a model for measurement error. More generally, Gauss proposed this theory of errors to argue that the **square** of error is the most optimal notion with which to work if one wishes to minimize error in scientific work. We know this concept as the **method of least squares** (the method of minimal squared error).

> Many scholars credit Laplace with the development of the normal distribution model for error, the fundamental premise of Gauss' Theory of Errors. Laplace, in fact, published the first formal treatise on the method of least squares. It is evident from his journals (diaries), however, that Gauss derived this method at age 17, predating Laplace's published work.

A significant application of Gauss' theory of errors is **regression analysis.** Regression literally means to "go back." In data analysis, regression refers to the tendency of the expected value of one of two correlated variables to move closer, go back, to the mean value than to any other value. Confusing? Explaining the choice of the word, regression, for the analysis of data is difficult. Many textbooks don't offer the explanation. Often, these books equate regression with the method of least squares. Although incorrect, the method of least squares is what most practitioners think of when the word, regression, is mentioned.

The term, **regress,** was used in place of **revert** by Galton (1886). In 1877, it was Galton who noted in a study of the mean diameter of offspring peas given the diameter of parent peas that the diameters of offspring peas tended toward the average offspring diameter. He felt these diameters reverted toward a common diameter, and later used the word, regress. When viewing these data in the following table, Galton observed that mean offspring diameter was greater than parent diameter for smaller parent diameters, and was lesser than parent diameter for larger parent diameters. The mean offspring diameter was 16.3, so offspring diameters tend more toward this quantity (regress toward it). The mean offspring diameter is consequently a pivot point for any model.

Galton's pea data (from Birkes and Dodge, 1993):

Parent Diameter (1/100 inches)	Mean Offspring Diameter (1/100 inches)
21	17.5
20	17.3
19	16.0
18	16.3
17	15.6
16	16.0
15	15.3

A plot of these data shows a linear relationship, with offspring diameters (Y) associated with a smaller range than parent diameters (X). It was this smaller range in Y that prompted Galton to think of these values reverting, or regressing, more toward a mean value in comparison to X.

In regression analysis, then, the expected value of Y is its mean value, the value that it regresses toward. Expected value was introduced in Chapter 2 in a brief sense in the section that describes the chi-square hypothesis test. The expected value of a variable, x, is written $E(x)$. Formally, $E(x)$ is that value, x, having the largest probability of occurring. If a normal distribution model is valid for a data distribution, then $E(x)$ is the mean value of x. Does the term, regression, imply normality? Not necessarily for the data, but errors are expected to be normal. If a variable, Y, is regressed on a variable, X, then the value, X, is used to predict Y. Calling the predicted value Y', then error is written as $Y - Y'$, where Y is the true value for X. The expected value, $E(Y')$, of the predicted values is equal to the mean value of Y. This requires the mean value of error, $Y - Y'$, to be zero, moreover that these errors be normally distributed. This is the essence of regression.

Gauss argued the normal distribution was a good model, a natural model, for errors of measurement. In regression, the application of this model is broadened to represent errors in general, even those due to fluctuation, whatever the reason for data variability. The main objective of regression analysis is **optimization.** This refers to finding a function, f, such that $Y' = f(X)$, that yields the smallest possible error. Moreover, the function, f, is a realistic model explaining the change in Y for any given change in X. In this sense, f is sought to provide physical insight to the process(es) that yielded X and Y. Gauss argued that the best way to minimize errors of regression is to use the square of error. If mean error is identically zero, then average squared error is equal to the variance of the error, or **estimation variance,** with the term, estimation, used as a synonym for prediction. Regression is now seen as a method for optimizing (minimizing) error variance.

3.1.1.1 Linear Least Squares Regression

This section is about best fitting a straight line to a cloud of points that are defined in x, y space. In this first treatment, x is assumed to be the **independent** variable. The implications of this designation are several and important. An independent variable is one that is known precisely, hence it has no error. It is the value that is varied in an experiment to observe an outcome. For instance, in a particular study of a newly developed car, its stopping distance versus speed of travel was assessed. In this case, velocity is the independent variable. In the present treatment, y is assumed to be **dependent** on x, such as stopping distance is dependent on rate of speed in the foregoing example. The dependent variable is assumed to be associated with error (measurement error). The normal distribution model is assumed to be valid for representing errors on y.

In **linear regression,** a line of the form, $y' = mx + b$, is assumed to represent the change in y as x changes. Is this a valid assumption? The only way to answer this question is by graphing the data (Figure 3.1). The most important first step in regression analysis is to create a graph of the data. The relationship between y and x is then visualized. If the data are highly variable (noisy) and seem to show little or no functional relationship, regression analysis should stop and the data be carefully reviewed and examined to determine the reason, or reasons for the noise. If the graph suggests a functional relationship, linear or nonlinear, regression analysis can proceed.

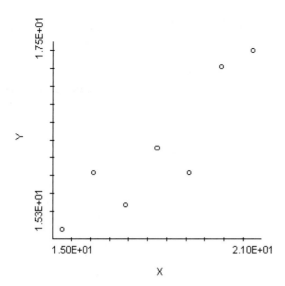

▲ FIGURE 3.1
Galton pea data. Mean offspring diameter (*Y*) is plotted versus parent diameter, *X*.

If y is assumed to carry all error, an expression for error can be written as $y - y'$, where, for any given value, x, y is the value of the dependent variable actually measured, and y' is that value yielded by the model, $mx + b$. Gauss argued that the square of the error is the best notion of error to use when attempting to optimize (minimize) it. Consequently, we can write error as $(y - y')^2$, which can be expanded as $y^2 - 2yy' + y'^2$. If we want the linear model to be a "best fit" over all data pairs, (x,y), then we write error as a total sum, $\sum(y^2 - 2yy' + y'^2)$, and this total sum is minimized to obtain the "best fit." This is the optimization.

To be precise, we wish to solve for the slope, m, and the intercept, b, with the y-axis such that error is minimum. This indicates that the minimum of the function for error must be found. Presently, though, m and b do not explicitly appear in the expression for error. Their presence is implied, however, by y'. Substituting for y' in the expression for error gives:

$$\text{error} = \sum\left(y^2 - 2mxy - 2by + 2mbx + m^2x^2 + b^2\right).$$

Differentiating this expression with respect to b, and setting this derivative to zero, yields the following solution: $b = \bar{y} - m\bar{x}$, where $\bar{y}$ and $\bar{x}$ are the mean values respectively of y and x. This solution indicates that, regardless of the value of m, the best fit line passes through the point, $(\bar{x}, \bar{y})$. Demonstrating this fact is a challenge to students at the end of this chapter.

It remains to solve for the slope, m, of the best fit line. Differentiating the expression for error with respect to m, and setting this derivative to zero, yields a solution for m that depends also on b. Substituting the solution for b into this expression yields an expression for the slope, m:

$$m = \left(n\sum xy - \sum x \sum y\right)/\left(n\sum x^2 - \left(\sum x\right)^2\right).$$

DEMONSTRATION

Given (x, y) data as follows: (1, 3) (2, 2) (3, 5) (4, 7) (5, 6). Find the slope, m, and intercept, b, for the best fit line. Notice that the solutions for both m and b are based on summations of x, y, xy, and x^2. If performing linear regression by hand, a convenient method to use is that of a table:

X	Y	XY	X²
1	3	3	1
2	2	4	4
3	5	15	9
4	7	28	16
5	6	30	25
Sum = 15	Sum = 23	Sum = 80	Sum = 55

In this demonstration, the data set size, n, is 5. The slope, m, is equal to $(5 * 80 - 15 * 23)/(5 * 55 - 15 * 15) = 1.1$. The intercept, b, is found using the solution for m: $b = (23/5) - (1.1 * 15/5) = 1.3$. The best fit line for these data is: $y' = 1.1x + 1.3$.

Once linear regression is finished to the point of solving for m and b, the regression model should be used to predict (estimate) the value, y', for each datum, x. Then, **residuals** (errors) should be computed for each datum, x, as $y - y'$. Furthermore, these residuals should be graphed versus x and examined for trends. Ideally, residuals should have a mean value of zero and be represented well by a normal distribution model. In addition to graphing the residuals, then, their histogram should also be computed and evaluated for normality. Any trend in the plot of the residuals versus x, or any deviation from normality as assessed by the histogram, indicates that a linear model is not representative of the change in y with x.

3.1.1.2 Linear Regression, Type II

What if both x and y have error? In other words, what if it is not possible to say which variable, x or y, is independent, and which is dependent? Such is the case when one wishes to understand the relationship, for instance, between gold and silver within a particular ore deposit; or the relationship between the visible green and visible red bands in the Landsat TM image of Walker Lake, Nevada. In these two examples (and numerous others), it is not even valid to think of these variables as independent or dependent. The method of linear regression presented in the preceding section should not be used to understand bivariate relationships when both variables are associated with error.

A different approach to linear regression is sought predicated on minimizing error on both x and y. We start again with the notion of error, but this time write it as a function of both x and y rather than just y. Using a visual image as a guide (Figure 3.2), we can write the following:

$$\text{error} = \sum\left(((y - y')^2 + (x - x')^2)^{1/2}\right)^2 = \sum\left((y - y')^2 + (x - x')^2\right).$$

Once again, substitution is necessary to obtain solutions for m and b. In this instance, $y' = mx + b$, the same linear model as is used in the preceding section. Consequently, $x' = ((y - b)/m)$. Substitution yields:

$$e = \sum_{i=1}^{N}\left(y^2 - 2mxy - 2by + 2mbx + \right.$$

$$\left. m^2x^2 + b^2 + x^2 - \frac{2xy}{m} + \frac{2bx}{m} - \frac{2by}{m^2} + \frac{y^2}{m^2} + \frac{b^2}{m^2}\right)$$

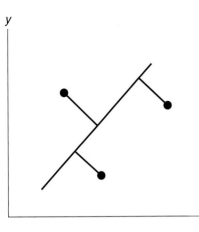

◀ FIGURE 3.2
The perpendicular distance from a point to
the best-fit line is minimized.

We realize that a fairly simple extension of least squares to account for error on both x and y has resulted in a substantially more complicated expression for total error, e. This expression is differentiated separately with respect to m, then b, to obtain solutions for m and b such that error, e, is a minimum. Kermack and Haldane (1950) was the first publication to present solutions for m and b for this type of linear regression. York (1966) presented a generalization to weighted regression. Solutions for m and b are:

$$m = \frac{\sum_{i=1}^{N} V_i^2 - \sum_{i=1}^{N} U_i^2 + \left[\left(\sum_{i=1}^{N} V_i^2 - \sum_{i=1}^{N} U_i^2\right)^2 + 4\left(\sum_{i=1}^{N} U_i V_i\right)^2\right]^{1/2}}{2 \sum_{i=1}^{N} U_i V_i}$$

$$\text{for which:} \quad U_i = x_i - \bar{x}; \quad V_i = y_i - \bar{y}$$

$$b = \bar{y} - m\bar{x}$$

[U is x minus mean of x; V is y minus mean of y; the intercept, b, is mean of y minus the product of slope, m, and the mean of x]

Notice that the equation for b for this type of regression matches that for the classical form of linear regression that accounts for error only on y. This implies that both forms of regression yield lines that pass through the point (mean of x, mean of y).

DEMONSTRATION

Given (x, y) data as follows: (1, 3) (2, 2) (3, 5) (4, 7) (5, 6). Find the slope, m, and intercept, b, for the best-fit line. Notice that the solutions for both m and b are based on summations of U^2, V^2, and UV:

X	Y	U (X − 3)	U²	V (y − 4.6)	V²	UV
1	3	−2.0	4.0	−1.6	2.56	3.2
2	2	−1.0	1.0	−2.6	6.76	2.6
3	5	0.0	0.0	0.4	0.16	0.0
4	7	1.0	1.0	2.4	5.76	2.4
5	6	2.0	4.0	1.4	1.96	2.8
			Sum = 10.0		Sum = 17.2	Sum = 11.0

From these calculations, we find that $m = (17.2 - 10 + \text{SQRT}((17.2 - 10)^2 + 4(11)^2))/(2(11)) = 1.38$. Using this value, b is found as mean, $y - m$ (mean, x) $= 4.6 - 1.38(3) = 0.46$. Notice that the slope, m, is greater than that which was determined by accounting for error only on y. Because both lines pass through the point, (3, 4.6), if slope, m, is greater, b must be less than that determined before. We see that the value, 0.46, is indeed less than the value, 1.3, determined before. The best-fit line, assuming equal amounts of error on x and y, is: $y' = 1.38x + 0.46$.

Comment: The line as defined by m and b above is called the "major axis." Kermack and Haldane (1950) went further to discuss standardizing x and y first by dividing by their respective standard deviations. From this, a simpler formula for slope, m, is obtained that is a function of the ratio, S_y/S_x. The slope is equal to this ratio only if the correlation coefficient (defined and discussed later in this chapter) is precisely equal to 1, or -1. This result yields what is known as the **reduced major axis.** If the level of error is approximately or precisely equal on x and y, York (1966) shows that the major axis solution is the preferred method of regression. This assumption is valid unless an analyst has a rationale for assigning unequal error to x and y. In this case, York (1966) derives the **least squares cubic,** a weighted regression method that is preferred when both x and y are associated with error, but the amounts of error differ.

3.1.2   On the Adequacy of the Linear Regression Model

Graphical displays of regression analysis (Figures 3.3 and 3.4) show how a model, in this case linear, is evaluated against the actual data. Earlier, it is stated that the essential beginning point of regression analysis is a graphical display to determine the nature of the relationship between two variables. Equally useful, a model can be selected by an analyst, applied to the data, and a graphical display developed to visualize the nature of the data relationship and how well the chosen model represents this relationship. In fact, the chosen model may help to refine the understanding of the bivariate relationship because if the model fit is visually poor it may be more obvious whether the relationship is linear or nonlinear. Of course, if

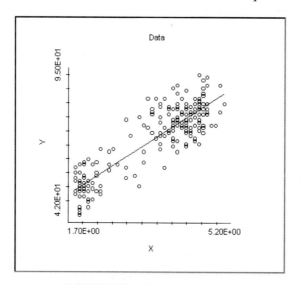

Slope, M = 10.35821

Intercept, B = 33.9667553020867

Correlation Coef, R = 0.877226884566493

F statistic for Slope = 734.556969591897

t statistic for R = 27.1027433573693

File: C:\text_cd\newdata\chap4\geyser.dat

▲ FIGURE 3.3
Linear regression applied to geyser eruption data (data are from Chatterjee, et. al., 1995).

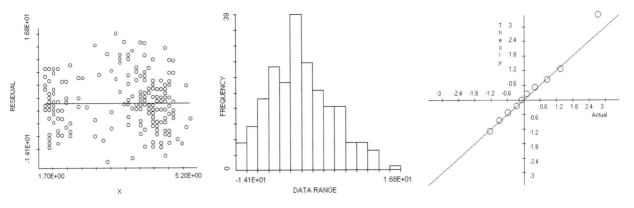

▲ FIGURE 3.4

Analysis of residuals from linear regression applied to the geyser eruption data. On the left is the plot of residuals. The center plot shows their histogram accompanied by the p-plot for the normal distribution model fit to these residuals. There are 222 residuals and the squared correlation coefficient for the p-plot is 0.98, indicating that a normal distribution model is representative of the distribution for these residuals.

the model represents a visually good fit, the nature of the bivariate relationship is confirmed.

Even if the visual fit between model and data appears to be good, a plot of residuals is necessary to confirm the fit (Figure 3.4). Recall that residuals are errors computed as $(y - y')$. These are plotted against x, even if regression is based on equal error for both x and y. In this case, residuals are the sine of total error. If a linear function is a good model, residuals should have a mean of zero and be normally distributed about this mean value. This necessitates a histogram analysis of residuals as part of the model validation process (Figure 3.4). The plot of residuals shows a horizontal line at $y = 0$; residuals should be equally distributed above and below this line (more or less equal proportions of negative and positive residuals) if the model fit is sufficient. Moreover, the residuals should plot as a cloud of points having a random appearance. Any noticeable pattern in the residuals suggests a deficiency in the chosen model for representing a bivariate relationship.

For example, Figure 3.5 is a contrived, artificial construct created using $y = 3e^{2x} + r$, where r is a random number. A **nonlinear** regression model is

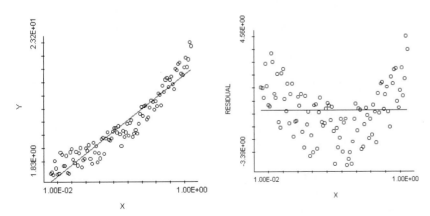

▲ FIGURE 3.5

Linear, least squares regression applied to untransformed artificial data. Notice that whereas the linear fit on the left appears to be good, the residuals show a distinctively nonrandom and nonlinear pattern, suggesting that the linear model is not adequate and a higher order model is needed.

necessary to adequately represent this relationship. Natural data often are the result of exponential processes for which a nonlinear regression model based on natural logarithms is necessarily applied. In this case, a model of the form, $1n(y) = 1n(b) + mx$, can be applied, enabling a linear regression of the data upon transformation of y to $1n(y)$. Applying this model to the artificial data yields a better fit (Figure 3.6). The value of the intercept, b, should ideally be equal to $1n(3)$ and is only a little less than this value due to the randomness, r, used in the artificial construct. The slope, m, should ideally equal 2, and is slightly larger than this value, again due to randomness in the artificial construct.

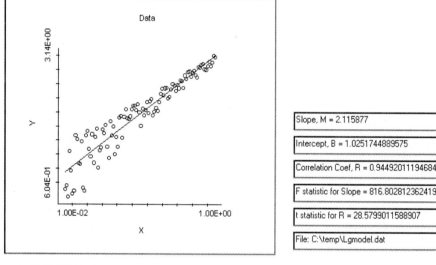

▲ FIGURE 3.6

A least squares, linear regression analysis of artificial data after transforming Y to $1n(Y)$. Notice the value of slope, 2.1, close to the ideal value, 2. Also notice the value for B, 1.025, close to the ideal value, $1n(3)$.

3.1.3 Hypothesis Testing of the Slope, *M*

In addition to visually inspecting plots of residuals, a formal assessment of the goodness of fit of a regression model is possible using statistical hypothesis testing. In this particular instance, we will test the size of the **error variance** against the size of the variance of the original data. Because we will be comparing variances, a hypothesis test known as the *F*-test is used.

3.1.3.1 The *F*-Test

Often, two sets of data are compared for numerical similarity. One way to compare numerical similarity is by comparing the means of the two sets. This is useful if **magnitude differential** is essential to the analysis. An example is when we wish to determine if the level of a contaminant in a sample of water exceeds acceptable limits. In this case, we must assess the difference of means against combined standard deviation; that is, data variability is also important. The type of hypothesis test used in this case is called a **t-test** and is introduced in detail later in this chapter.

Another way to compare the numerical similarity between two sets of data is by computing a ratio of their variances. This is useful when data are the result of some process, or processes, that can be altered. A manufacturing process, for instance, may be modified in an attempt to improve product quality. Suppose size is a measure of quality. Any reduction in size variation is an indication that the modification had the desired effect. In this case, computing a ratio of larger variance over smaller variance is the metric used in the type of hypothesis testing known as the **F-test.**

In regression analysis, the objective is to minimize total, cumulative squared error. If mean error is zero, squared error represents **error variance.** The size of error variance as measured against original variance is a measure of the quality of the regression. Recall that regression literally means to "go back." In data analysis, this refers to the data being regressed (usually y) having a tendency more toward the mean value of y ("going back to the mean;" or, "moving toward the mean"). The closer values are to the mean, the smaller variance will be.

Using this notion, the following calculations are used to implement an F-test for regression analysis. First, the hypotheses are stated as follows:

H_0: regression variance = Standard error variance

H_a: regression variance is not equal to Standard error variance.

If a regression model is not sufficient for explaining the variation in y with changes in x, the regression variance and error variance will be more alike. The greater the regression variance compared to the error variance, the more significant is the regression (the better is the model). Variance of the regression is computed by calculating the variance of the predicted values, y'. Error variance is standardized (divided) by $N - 2$. Then, a statistic, F, is computed as

$$F = (\text{regression variance})/(\text{standardized error variance})$$

The value of this statistic is compared to tabulated values of the **F-statistic** (Appendix C) using 1 numerator degree of freedom and $N - 2$ denominator degrees of freedom. If the computed F value is greater than the tabulated value, the null hypothesis is rejected and the regression is determined to be significant.

DEMONSTRATION

Given the problem used earlier for regression assuming error is on y only, for which $y' = 1.1x + 1.3$, then

X	Y	Y'	Y − Y'
1	3	2.4	0.6
2	2	3.5	−1.5
3	5	4.6	0.4
4	7	5.7	1.3
5	6	6.8	−0.8
Sum = 15	Sum = 23	variance = 3.025	variance = 1.275

Standardized squared error is obtained by dividing total squared error by $N - 2$; in this case, standardized squared error equals 1.275/3, or 0.425. The F-statistic is computed as 3.025/0.425 = 7.12. The table value (Appendix C) for the F-statistic for 1 numerator degrees of freedom and 3 denominator degrees of freedom, assuming a level of significance of 0.05, is 10.13. We therefore DO NOT reject the null hypothesis and may not conclude that this regression is significant.

What about the results for the other method of regression that assumes equal error on x and y? From that method, we obtained a model, $y' = 1.38x + 0.46$. If we apply this model to the data, then regression variance is 4.761, and standardized error variance is 0.49, yielding an F-statistic of 9.72. Still, we cannot reject the null hypothesis, but the regression model assuming equal error on x and y is a better fit.

3.1.4   Correlation Coefficient

Regression analysis helps us understand, particularly by visual inspection, the relationship between two variables. The word, relationship, has been used liberally thus far in this chapter. A new term will now be used in its place. **Correlation** is a measure of the goodness of fit of a **linear** model to bivariate data. Correlation is described numerically by the **correlation coefficient, r**. This coefficient is related to the slope of the linear regression model as $m = rS_y/S_x$, which holds that correlation coefficient, r, is computed from the slope of the regression model as $r = mS_x/S_y$. Correlation will now be used in place of the term, relationship.

The equation for r is valid in the case for which x is truly independent and error is on y only. In the case where x and y are both associated with equal error, the following formula is used for r; Kermack and Haldane (1950):

$$r = \frac{\sum\limits_{i=1}^{N} U_i V_i}{\sqrt{\sum\limits_{i=1}^{N} U_i^2 \sum\limits_{i=1}^{N} V_i^2}}; \quad U_i = x_i - \bar{x}; \quad V_i = y_i - \bar{y}$$

Using this form of the correlation coefficient, the reduced major axis slope may be computed as; Kermack and Haldane (1950):

$$m = \frac{S_y}{S_x}\left[1 \pm \sqrt{\frac{1 - r^2}{N}}\right]$$

Correlation coefficient, from either method, ranges in value from -1 to 1. A value, $r = 1$, represents perfect, positive correlation. Y changes in the same way, positive or negative, that x changes. A value, $r = -1$, represents perfect, negative correlation. Y changes in the opposite way that x changes; if x increases, y decreases; if x decreases, y increases. A value, $r = 0$, represents no correlation, or **independence** between the two variables. Values of r represent the degree of correlation. The closer r is to 1, the better is the positive correlation. The closer r is to -1, the better is the negative correlation.

A more severe metric for assessing correlation is the square of the correlation coefficient, r^2. A value, $r = 0.6$, may seem like good correlation, yet r^2 is equal to 0.36, which doesn't seem as good. A value, $r = -6$, has the same r^2 value, 0.36. A useful rule is that "good" correlation is inferred if r^2 exceeds 0.5. There is no formal basis to this rule and its subjectivity is noted.

Using the same data set analyzed earlier, the standard deviation of x, S_x, is 2.5 and that for y, S_y, is 3. Correlation coefficients are determined as follows:

1. assuming all error is on y: $r = mS_x/S_y = 1.1(2.5)/3 = 0.92$
2. assuming error is equally on both x and y, then $r = 11/\text{SQRT}(172) = 0.84$
3. using the value, $r = 0.84$, the slope m, of the reduced major axis is $(3/2.5)(1 + .24) = 1.49$. This example is introduced to show that the reduced major axis approach yields yet a different line. The major axis, the one for which $m = 1.38$, is preferred over the reduced major axis approach.

Notice the difference in correlation coefficients depending on the assumption, and why it is so important to consider the nature of the data, x and y, and whether error is on only y, or both x and y.

3.1.5   Covariance

Correlation is also a function of the **covariance** between two variables. In the preceding chapter, variance was introduced as the second statistical moment, a metric for data variability. In bivariate data analysis, the notion of how the variables vary with respect to one another is important. In equation form, covariance may be computed as

$$COV(x,y) = \frac{1}{N-1} \sum_{i=1}^{N} (x_i - \bar{x})(y_i - \bar{y})$$

[Covariance is equal to the cumulative product between (x minus its mean) and (y minus its mean) divided by the number of degrees of freedom, $N - 1$]

Another way to compute covariance is as follows

$$COV(x,y) = \frac{N \sum_{i=1}^{N} (x_i y_i) - \sum_{i=1}^{N} x_i \sum_{i=1}^{N} y_i}{N(N-1)}$$

From this expression for covariance, we realize that the slope, m, in linear regression (in which all error is on y) can be computed as $COV(x, y)/VAR(x)$, in which $VAR(x)$ is the statistical variance of x. Moreover, an expression for correlation coefficient may be written as $r = COV(x, y)/\text{SQRT}(S_x S_y)$. Correlation coefficient, r, is directly proportional to covariance.

3.1.6   On the Statistical Significance of the Correlation Coefficient

Hypothesis testing is possible to determine the statistical significance of the correlation coefficient, r. In this instance, the following hypotheses are tested:

$$H_0: r = 0; \qquad H_a: r \text{ is not equal to zero.}$$

Because we are essentially comparing two values, r and zero, a t-test is a useful form of hypothesis testing. Such a test is used to compare two values, subject to data variability. One form of a t-test compares the mean (average) of a set of data to some "ideal" value. For instance, N samples of soil were taken at a site suspected of being contaminated by lead. The average lead concentration is found to equal W parts per million, with a standard deviation of Y parts per million. The United States Environmental Protection Agency's maximum allowable lead concentration is *Pb-max*. Is there statistically significant evidence that the site is contaminated by lead?

A t statistic is computed as

$$t = \frac{W - Pb_{MAX}}{\frac{Y}{\sqrt{N}}}$$

The purpose of this illustrative problem is to show that a t statistic is a metric for determining the significance of the difference between two values, in this case the mean of a set of data and some comparative value. The significance of t is determined by comparing its absolute value to table values (Appendix D) for $N - 2$ degrees of freedom. Typically, a 0.05 level of significance is used.

In the case of correlation coefficient, a t statistic is computed as follows:

$$t = \frac{\frac{r - 0}{\sqrt{1 - r^2}}}{\sqrt{N - 2}} = \frac{r\sqrt{N - 2}}{\sqrt{1 - r^2}}$$

Notice that this is the same equation as used in the previous example. Because r is compared to zero some simplifications of the equation are possible. As in the previous example, the absolute value of t is compared to table values (Appendix D) for $N - 2$ degrees of freedom and a level of significance of 0.05.

3.2   Visual Regression

3.2.1   Linear Models

Blind acceptance of a least squares fit of a model to data without visually inspecting both the fit and the residuals is poor scientific practice. In fact, the mathematics represents a tool for aiding the visualization. The ultimate endpoint of regression is the visualization.

Some examples illustrate potential pitfalls of regression analysis if visualization is ignored.

`Case 1` A single data point disproportionately influences linear, least squares regression:

Consider the following bivariate data set:

X	Y	X	Y
1.0	23.1	2.0	20.4
3.0	19.1	4.0	15.6
5.0	12.9	6.0	20.1
7.0	10.5	8.0	11.3
9.0	7.2	10.0	8.1

A visual inspection afforded by a two-dimensional plot shows a decreasing trend:

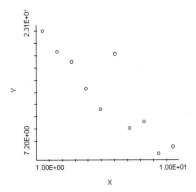

One unusual point defined by $X = 6.0$ and $Y = 20.1$ is identified. Applying least squares linear regression to these data yields the equation, $y = 24 - 1.66x$, with an F-statistic for the fit of 32.55. A plot of residuals reveals the substantial error on the one unusual point:

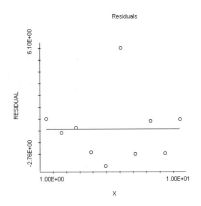

Suppose the one unusual datum, $X = 6.0$, $Y = 20.1$, is removed from the set. Then, least squares linear regression yields the model, $y = 23.5 - 1.7x$, with an F-statistic for the fit of 114.44. The model seems not to have changed significantly, but the F-statistic suggests otherwise.

3.2.1.1   Least Absolute Deviation (LAD) Regression

A method of regression that is less sensitive to unusual, or extreme data values is based on minimizing the absolute value of error, rather than squared error. Whereas simpler in mathematical representation, this method is more difficult to solve compared to the method of least squares. **Least absolute deviation (LAD)** fitting of a line to a cloud of points seeks to fit a model, $y = mx + b$, to bivariate data, $x - y$, such that total, absolute error is minimized. As with the other forms of linear regression, the intercept, b, is equal to the mean value of y, less the product of slope, m, and the mean value of x. Finding the slope, m, however is somewhat complicated with this method.

First, a ratio is computed for each bivariate pair as $(y_i - \text{mean}_y)/(x_i - \text{mean}_x)$. Second, these ratio values are sorted in order, smallest to largest (ascending order). Third, using these sorted ratios, the total sum of their denominators is computed: $S = \sum Abs(x_i - \text{mean}_x)$. Fourth, the subscript, J, is found such that $Abs(x_{J-1} - \text{mean}_x) < 0.5S < Abs(x_J - \text{mean}_x)$. Finally, slope, M, is computed as the ratio, $(y_J - \text{mean}_y)/(x_J - \text{mean}_x)$.

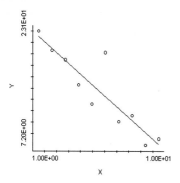

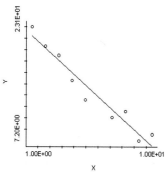

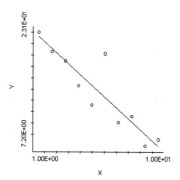

▲ FIGURE 3.7

Comparison of three linear regression models: top from least squares applied to the full data set; middle from least squares applied to the filtered data set; and bottom from least absolute deviations applied to the full data set.

Applying least absolute deviation regression to the complete data set of Section 3.2.1, including the one unusual datum, $X = 6.0$, $Y = 20.1$, yields a linear model, $y = 24.2 - 1.71x$ with an F statistic of 34.21. The slope in this instance is similar to that obtained when this one unusual datum was removed. The significance of the least absolute deviation regression is similar to that for linear, least squares regression applied to the full data set that includes the one unusual datum. Applying least absolute deviation regression to data associated with extreme values is a better alternative to filtering because all available data are included in the analysis. The slope obtained using least absolute deviation regression is insensitive to these extreme values. All three regression experiments applied to the data of Section 3.2.1 are shown on the left for comparison.

3.2.1.2   Comment: Least Squares or Least Absolute Deviation, Which Should be Used?

Provided there are no unusual or outlying points, least squares is usually a better performer when compared to least absolute deviation (LAD), provided residuals are normally distributed. If there are unusual points in a data set, LAD is often the better algorithm to use. Moreover, LAD is a more robust method. In statistical analysis, **robustness** refers to sensitivity of the analysis to a distribution model, particularly normality. A robust method is one that is relatively insensitive to data distribution, whereas a nonrobust method is sensitive to data distribution. Because of its premise that residuals are normally distributed, least squares is a nonrobust method. LAD is more robust.

Is there a form of regression that can use least squares regression when it is optimal, or LAD when it is optimal, or automatically decide which to use? Yes, this method of regression is referred to as **M-Regression.** Its algorithm is fairly simple:

1. Apply linear, least squares regression to the data and determine slope, M, and intercept, b;
2. For each of the N data values, compute absolute deviation as $Abs(y_i - Mx_i - b)$;
3. Compute the median of these absolute deviations (MAD); multiply this value by 1.483 to obtain an estimate of the population standard deviation;
4. For each of N data values, if its absolute deviation is greater than 1.483MAD, then change y_i to $y_i = Mx_i + b + e$, where e is the adjusted deviation, -1.483MAD if the actual deviation for the ith datum is negative, or 1.483MAD if the actual deviation is positive. All N data locations are evaluated in this manner, and if necessary y is adjusted.
5. Step 1 is repeated, but with the adjusted values, y, and new values of M and b are computed using least squares. Then, steps 2 through 4 are completed. This process iterates until M and b no longer change. The final values of M and b are those used in the regression analysis.

If there are no unusual data values present in the data set, M-regression is identical to linear, least squares regression. If there are unusual data values present, then M-regression is similar to LAD. The advantage of M-regression is its ability to switch between these two algorithms automatically to find an optimal fit of a model, linear in this case, to bivariate data.

3.2.2   Analysis of Residuals to Infer the Need for Nonlinear Regression

The following bivariate data represent spectral response in the visible green and red portions of the electromagnetic spectrum of Martian sky imaged by the Mars Pathfinder camera:

X (green)	Y (red)	X (green)	Y (red)
175	154	175	155
174	155	172	154
169	152	175	153
175	154	174	154
172	153	169	151
175	152	175	153
174	154	172	153
169	150	175	152
175	153	174	154
172	153	169	151

Application of linear, least squares regression to these data yields residuals as shown in Figure 3.8. Not only do the residuals plot more on the positive side of the zero error line, their distribution in this plot does not appear to be random. One way to assess the goodness of fit of a regression model is to visually inspect the plot of residuals for randomness and distribution around the zero error line. A good fit results in almost equal distribution of residuals on the positive and negative sides of the zero error line and the pattern of residual plot appears random. If not random, then there is still some function that can be fit to the residuals. Whatever this function is, it should be incorporated into the original regression model. Typically, if the plot of residuals from linear regression does not appear random, then some higher order regression model is warranted.

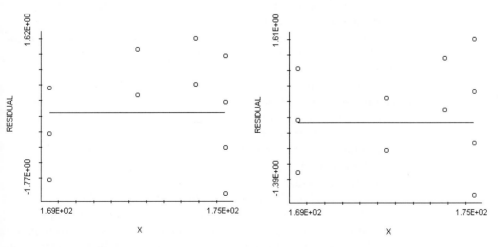

▲ FIGURE 3.8
On the left, residuals from linear regression do not plot equally above and below the zero error line. Moreover, there appears to be a nonrandom pattern to the residuals. On the right, application of seven order, nonlinear regression yields residuals almost equally distributed above and below the zero error line.

One form of nonlinear regression, that which is based on converting one or both bivariate data pairs to their natural logarithms, was introduced earlier in this chapter. In general, regression models of order N are possible:

$$y = B_0 + B_1 x + B_2 x^2 + \cdots + B_N x^N$$

The key is finding the order, N, for which the plot of residuals is most nearly optimal. In the case of the Mars Pathfinder data, a seventh order model is found to yield this optimal plot. Actual regressions are shown below.

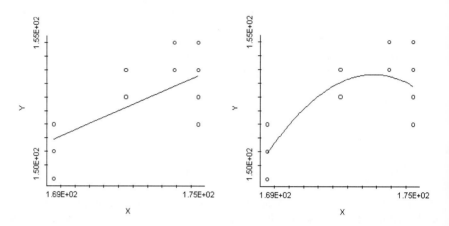

▲ FIGURE 3.9
On the left, a linear model is fitted to the Pathfinder green (X) and red (Y) reflectance. On the right, a seventh order model is fitted to reflectance and yields residuals more evenly distributed around the zero error line.

Is a seventh order model scientifically meaningful for these data? There are many different models that can be fit to a set of bivariate data. Few, if any are valid, where validity is assessed statistically, by using hypothesis testing, visually by viewing the fit of the model and judging correctness of the model, or scientifically, whereby the order of the model describes the perceived functional relationship between x and y.

In this particular case, the seventh order model is associated with a decrease in red reflectance for higher green (X) reflectance. Is this realistic? The data set is fairly small, twenty data pairs in size. Is this data set representative of the whole (the total image represents approximately 80,000 data pairs)? These are important questions to consider when evaluating the results from any regression analysis, linear or otherwise.

A larger extraction of green and red reflectance data from this image is used to develop the plot on the following page.

This plot visually suggests that a linear relationship exists between green (X) and red (Y) reflectance.

The application of nonlinear regression to the smaller data set is now seen to serve two purposes. It represents a simple demonstration on how to implement a nonlinear regression and how to use residuals to determine an optimal fit. Yet, there is a substantial burden on the analyst to justify the higher order model for modeling a bivariate relationship. In the particular application to Pathfinder reflectance, the higher order model is not justifiable based on visual inspection of a larger sample.

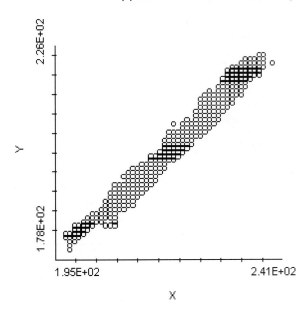

▲ FIGURE 3.10
A bivariate plot (scatter diagram) reveals a linear trend between green and red reflectance.

3.3 Application of Correlation Analysis to the *Nevada_Landsat* Data Set

Visual_Data is applied to the *Nevada_Landsat* data set that is supplied on your CD-ROM in the directory \data\nvlandat\. Based on the histogram analyses presented in Chapter 2 (Table 2.1), we infer that none of the data for spectral bands comprising the *Nevada_Landsat_6x_(nvlan6x)_data* set are represented well by a normal distribution model. The assumption inherent to linear, least squares regression is that the **residuals** from the model are normal. Does this necessitate normality in the data?

Not necessarily. What is quite important in this particular analysis is the question of error on X and Y. It is not valid to designate any of the spectral bands as the independent (or dependent) variable. The spectral bands have approximately equal error (this may not be true for the thermal band; this particular band represents coarser resolution compared to the other six bands, and the sensing instrument is different; nevertheless, we will assume it has error in an amount equal to the other bands). Given that X and Y are both associated with error, the method of regression chosen for this analysis is linear, least squares type II that minimizes perpendicular distances between the model line and data points.

Seven (7) variables comprise the *Nevada_Landsat_6x* data set. If we choose two of these at a time to form bivariate combinations, we find that there are $(7 \times (7 - 1))/2$, or 21 different bivariate combinations. This is too many to show in this section, one that is intended simply to illustrate concepts. By arbitrary choice, we will look at the following bivariate pairs: (blue and green); (blue and red); (blue and near infrared); (green and red); (green and near infrared); and (red and near infrared). Moreover, simply as an experiment, the following two pairs will be examined: (red and thermal); and (near infrared and thermal). In total, we will apply regression to eight bivariate combinations. Graphs of these regressions are shown in Figure 3.11.

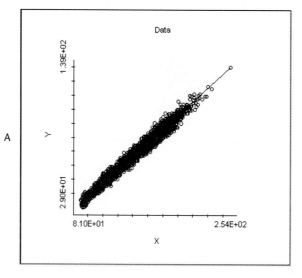

A

Slope, M = 0.6183074

Intercept, B = -18.3420215725183

Correlation Coef, R = 0.988384921988249

F statistic for Slope = 427661.120769485

t statistic for R = 650.311421452647

File: C:\Visual_Data\Walker_Lake_Data\Wal

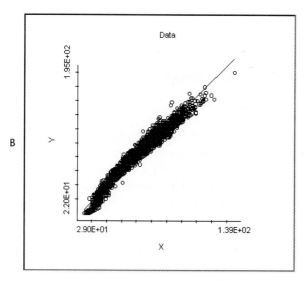

B

Slope, M = 1.686787

Intercept, B = -22.0200460190892

Correlation Coef, R = 0.982995133255211

F statistic for Slope = 295806.650758301

t statistic for R = 535.254704602795

File: C:\Visual_Data\Walker_Lake_Data\Wal

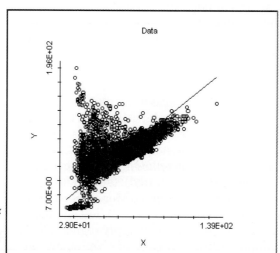

C

Slope, M = 1.50129

Intercept, B = -23.8412539488196

Correlation Coef, R = 0.726074956531729

F statistic for Slope = 19901.4785686578

t statistic for R = 105.582594155234

File: C:\Visual_Data\Walker_Lake_Data\Wal

▶ **FIGURE 3.11**
Application of linear regression, modeling equal error on x and y, for the following bivariate combinations from the *Nevada_Landsat_6x* data set: A. visible blue (x) and visible green (y); B. visible blue (x) and visible red (y); C. visible blue (x) and near infrared (y);

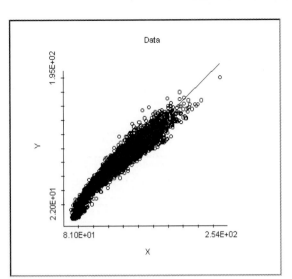

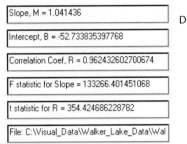

Slope, M = 1.041436

Intercept, B = -52.733835397768

Correlation Coef, R = 0.962432602700674

F statistic for Slope = 133266.401451068

t statistic for R = 354.424686228782

File: C:\Visual_Data\Walker_Lake_Data\Wal

D

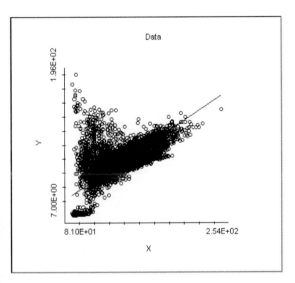

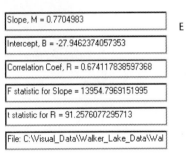

Slope, M = 0.7704983

Intercept, B = -27.9462374057353

Correlation Coef, R = 0.674117838597368

F statistic for Slope = 13954.7969151995

t statistic for R = 91.2576077295713

File: C:\Visual_Data\Walker_Lake_Data\Wal

E

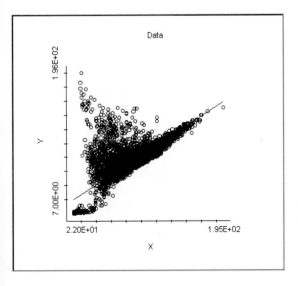

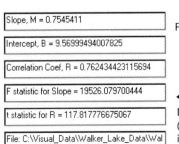

Slope, M = 0.7545411

Intercept, B = 9.56999494007825

Correlation Coef, R = 0.762434423115694

F statistic for Slope = 19526.079700444

t statistic for R = 117.817776675067

File: C:\Visual_Data\Walker_Lake_Data\Wal

F

◀ FIGURE 3.11 (*continued*)
D. visible green (*x*) and visible red
(*y*); E. visible green (*x*) and near
infrared (*y*); F. visible red (*x*) and
near infrared (*y*);

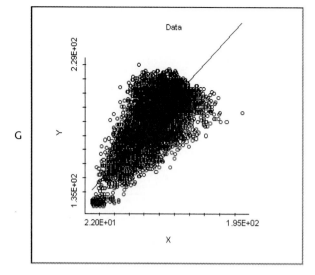

G

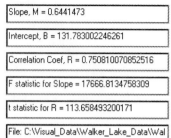

Slope, M = 0.6441473

Intercept, B = 131.783002246261

Correlation Coef, R = 0.750810070852516

F statistic for Slope = 17666.8134758309

t statistic for R = 113.658493200171

File: C:\Visual_Data\Walker_Lake_Data\Wal

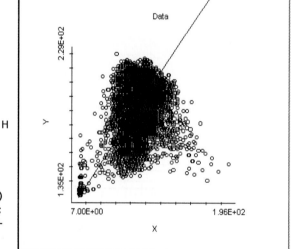

H

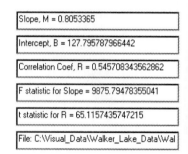

Slope, M = 0.8053365

Intercept, B = 127.795787966442

Correlation Coef, R = 0.545708343562862

F statistic for Slope = 9875.79478355041

t statistic for R = 65.1157435747215

File: C:\Visual_Data\Walker_Lake_Data\Wal

▶ FIGURE 3.11 (*continued*)
G. visible red (*x*) and thermal (*y*);
and H. near infrared (*x*) and thermal (*y*).

What can we make of these weird outcomes? Based on visual appraisal, a linear model is appropriate only for explaining the variation in visible blue reflectance with visible green reflectance (Figure 3.11, A). The relationship between visible blue and visible red (Figure 3.11. B) and between visible green and visible red (Figure 3.11.D) show slightly nonlinear behavior. The relationship between visible red and thermal (Figure 3.11.G) appears to be linear, but a plot of residuals shows that the linear model is deficient. The remaining bivariate plots reveal complicated and nonlinear relationships for bivariate pairs.

Figure 3.12 shows that residuals from the linear regression of visible green on visible blue (Figure 3.12, A) are fairly randomly distributed in more or less equal proportions above and below the zero residual line. The other three plots of residuals, however, reveal nonrandom patterns, suggesting a deficiency in the linear model. The regressions of visible red on visible blue, and visible red on visible green suggest that a nonlinear model is more appropriately applied for each. The regression of thermal on visible red likewise suggests that the linear model is deficient and that a higher order model is warranted.

In one experiment, linear regression of visible red on visible blue is attempted once again. But, the visible blue values are first transformed to their

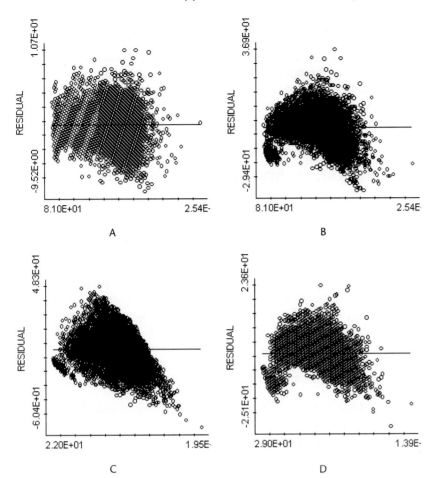

▲ FIGURE 3.12
Residuals: A. blue and green; B. blue and red; C. green and red; and D. red and thermal.

natural logarithms. The resulting model fit and residual plot are shown in Figure 3.13. In this instance, the nonlinear transform applied to visible blue values allows a linear model to be more representative of the variation in visible blue with visible red. This is an interesting outcome that was realized through ex-

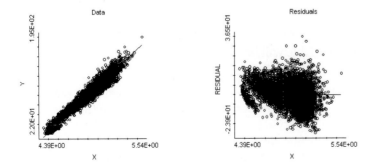

▲ FIGURE 3.13
Applying a log transform only to visible blue *(x)* yields the bivariate plot shown on the left. A linear model appears to be representative of the variation between $1n$(blue) and red. The residuals on the right still show some nonrandomness, yet do appear to be more random than what is obtained when no log transform is applied to visible blue (please compare to Figure 3.12, B).

perimentation by first attempting log transforms of both visible blue and visible red, then only of visible red, and finally only of visible blue. In the first two transform experiments, the linear model was not representative of the bivariate plot. Results of the third experiment are shown in Figure 3.13. This seems to suggest that a nonlinear relationship exists between visible blue and visible red. This is an interesting realization, especially given the speculation presented earlier in this chapter regarding results from the Mars Pathfinder camera.

3.4   How Do I Reproduce Results in This Chapter, or Analyze My Own Data . . .

3.4.1   . . . Using *Visual_Data?*

Once the program is started and the main window is showing, click File, then Open to access an existing file, or click File, then New to create a file. For regression analysis, from one to an unlimited number of variables can be accommodated per sample. If only one variable is on a line, then this variable is considered by the program to be Y, moreover X is considered to be implicit and to represent regular sampling. In the case where only one value occurs per line, *Visual_Data* will prompt interactively for information about X. If creating a data set using New, then save the file when finished entering data by clicking File, then choosing Save (not Save As).

Once a data file is opened, click Tools, move the mouse pointer to Regression, and the following options will appear:

- Linear, least squares, classical (note that error is assumed for y only and x is precise);
- Linear, least squares Type II, with equal error on x and y;
- Least absolute deviation (LAD) linear regression
- M-regression, linear model (limited to no more than 5 iterations);
- Nonlinear, bivariate regression
- Multivariate, linear regression (see Chapter 4)

Choose the option that you wish to apply. If you are unsure, use the Two-Dimensional, X-Y Plotting tool to graph a scatterdiagram. Look at this plot to decide what type of function may be representative of the relationship between your bivariate data. Also, think about error and whether it is all on Y, or equally on X and Y. Finally, choose a regression option for experimentation. The first five options will ask for the following information interactively:

- How many variables are on each line of your data file?
- Which variable will be considered as X.
- Which variable will be considered as Y.
- Do you wish X and/or Y converted to their natural logarithms?
- Finally, a user is prompted to save residuals to a file.

Additionally, the nonlinear, bivariate option asks a user to select the order of the regression, two or higher, before seeking responses to these five interactive questions.

Two plots are produced automatically by the regression tool. One is a plot of residuals. Use this plot to determine the adequacy of the chosen regression model. This form may be printed by clicking the Options button. Once finished with this form, click Options, then Close this form. This will display the main regression window showing a plot of the regression, with chosen model, and several text boxes that show slope, intercept, and results from the statistical hypothesis tests. Clicking Options will also reveal a printing option for this form. When finished with this form, click Options, then click Return to the Main Program.

3.4.2   . . . Using MATLAB 5.3?

MATLAB is capable of performing all forms of regression described in this chapter, provided the user programs the steps. For the first example, the function, xyplot, coded and described by Middleton (2000, p. 60) is applied to the data set, geyser.dat. The plot of the regression results, followed by analytical results, are shown below:

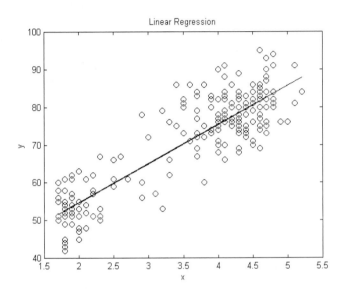

```
Analysis of Variance Table.

Source          d.f.      Sums of Squares  Mean Squares
-----------------------------------------------------------
Regression      1         27859.93         27859.93
Residuals       220        8344.06            37.93
-----------------------------------------------------------
Total           221       36203.98

F = 734.557
Correlation coeff: r = 0.877227
Equation: y = 33.9668 + 10.3582 x

Click on Window to see figure

ans =

   10.3582    33.9668

>> |
```

These results match those shown earlier (Figure 3.3). Applying two, independent programs for data analysis helps to verify results from each.

This MATLAB result was obtained using the following MATLAB commands:

```
>>load geyser.dat
>>X = geyser
```

(Middleton's book comes with a 3.5 inch, high density diskette with *MATLAB* functions; this disk was placed in the A drive):

```
>>cd a:
>>xyplot (X(:,1), X(:,2))
```

This last command resulted in the plots shown above.

The function, xyplot, yields a least squares linear regression in which all error is assumed on y and x to be precise. In the type II regression wherein equal error is assumed on x and y, the function, lreg (Middleton, 2000, p. 61) is modified and renamed, lreg_t2, and is found on the CD-ROM at the end of this text in the folder, MATLAB. The application of this function to the geyser eruption data (CD-Rom:\data\othrdata\geyser.dat) yields the following results:

```
ans =
    22.9517     13.4384
```

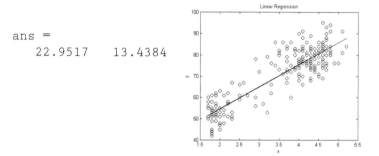

These results match those from *Visual_Data* (not shown).

The function, lreg, is also modified to perform *M*-regression. This method is the ideal combination of least absolute deviation regression and classical least squares. The modified function, mreg, is also found in the MATLAB folder. The application of this function to the geyser eruption data yields the following results:

```
ans =
    33.7832
    10.3607
```

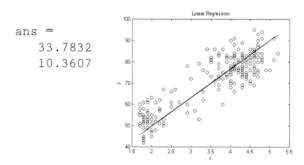

These results match those from *Visual_Data* (not shown).

3.4.3   . . . Using *Microsoft Excel?*

Microsoft Excel offers a powerful tool for classical least squares linear regression. Data, such as the Galton pea data, are loaded into *Excel* as shown:

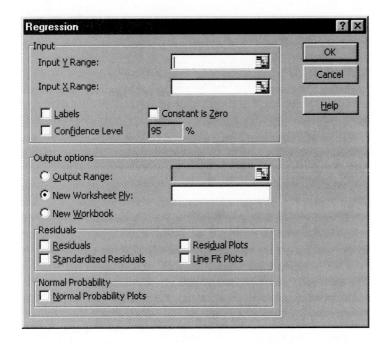

Once the data are loaded, click Tools, then Data Analysis, then highlight Regression and click OK. The following window appears:

Using *Excel* is a bit tricky at this point, and its Help capability is rather poor in explaining how to use this regression window. Two important categories are shown at the top of this window: Input Y Range, and Input X Range. Unlike Excel's histogram tool, in which the range can be easily specified by clicking the column letter for the data to be analyzed, the regression tool requires a more specific response. The easiest way to specify input X and Y ranges is to click the small icon containing a small, red arrow located on the right side of the X and Y input range windows. This will allow you to use your mouse to highlight all data values that you wish to be considered as Y or X. For instance, the regression window is shown below for the Galton pea data to show how the input ranges are to be specified, and how other options were set.

The form of regression yielded by *Excel* is classical linear regression. Performing the type of regression that accounts for equal error on X and Y, or some other form of regression such as M-regression, requires inserting functions for specific calculations of the different components of the formulas necessary to the calculation of slope and intercept. An example spread sheet for type II regression applied to the Galton pea data is written on the CD-ROM in the folder, *Excel*, with the file name, Galton.xls.

3.5   Weighted Regression

Suppose error on Y, or joint error on X and Y, is not constant. That is, some data values are associated with more "confidence" than others. If this is the case, then those values associated with higher confidence may be given greater weight (may be allowed to more greatly influence) in regression.

This is accomplished by multiplying each data value by some number, called a **weight.** The numerical size of the weight establishes the influence a data value is allowed in regression. For instance, filtering a datum is tantamount to weighting the datum by zero. If error is on Y only, this error increases with X, then the weight for any given value, Y, might be $1/X$. This is only an example. There are an unlimited number of possible weighting schemes. For instance, if the influence of one value, Y, in regression is forced to be twice that of another, then the weight on the more influential value is set equal to two (that value is multiplied by two). This is simply another example to illustrate the concept of unequal weighting.

Weights may be used in any of the regression methods that are described in this chapter. As implemented in this chapter, the programs, *Visual_Data*, the functions used in MATLAB, or *Microsoft Excel* cannot explicitly accommodate weights. Instead, a user performs weighted regression by premultiplying data by weights to yield new data values. Then, a method of regression is chosen with one of these three programs to obtain results.

3.6   A Review of the *Nevada_Landsat* Data Set

The *Nevada_Landsat_6x* data set is now analyzed through two chapters. From Chapter 2, we learned that none of the seven variables associated with this data set has a distribution that is modeled well by the normal distribution. In this chapter, we learned that only two of the seven variables, visible blue and visible green, have a bivariate relationship that is truly linear. Other bivariate combinations from this group of seven variables show more complicated relationships. The variables representing visible light, blue, green, and red, are more linearly related with one another than they are with the infrared or thermal variables. Consequently, there is more intercorrelation among frequencies of visible light, and correlation is lesser between a particular visible frequency and near infrared, mid-infrared, and thermal frequencies.

3.7   Literature

The book by Birkes and Dodge (1993) was inspirational for much of this chapter. York (1966) likewise inspired much of the writing of this chapter from the perspective of accommodating equal error on X and Y, a rather common situation faced by researchers and students of the natural sciences, geology, biology, meteorology, hydrology, and hydrogeology. A motivational work was that on regression graphics by Cook (1998). Finally, Middleton (2000) was not only helpful when second guessing statements forwarded in this chapter, but was especially useful for understanding how to use MATLAB for regression analysis.

Exercises

Problems in this section may be solved by hand, with *Visual_Data*, MATLAB, or *Microsoft Excel*.

1. Compute by hand the least squares linear regression of the following data:
   a. X: 10 9 8 7 6
      Y: 1 3 2 5 4
   b. X: −2 −1 0 1 2
      Y: 9   8 9 8 9
   c. X: 1   2   3   4   5
      Y: 0.5 5 11   15   30
   d. X: 1   2   3   4   5
      Y: 12. 200 900 11000 200000
   e. X: 2 4   6   8   10
      Y: 4 8 12 40   20

2. For which data sets in Problem 1, a–e, is a linear model not appropriate?

3. For which data sets in Problem 1, a–e, is a linear model appropriate upon transforming Y to base-10 logarithms?

4. For which data sets in Problem 1, a–e, is LAD regression a better approach?

5. Show for Type II regression that the intercept, b, is equal to mean $(y) - m(\text{mean}(x))$.

6. Analyze the data of Problem 1 using one of the computer programs described in this chapter. Do these computer results verify your hand calculations?

7. By hand, compute the covariance and correlation coefficients for the five data sets in Problem 1. How do your hand calculated correlation coefficients compare to the computer results obtained in Problem 6?

8. Use one of the three computer programs to repeat the analysis of the *Nevada_Landsat_6x data*. Do your results match those shown in this chapter?

9. Derive the equations for m and b in classical, linear, least squares regression. Show that this line passes through the point defined by the mean of x and the mean of y.

10. The CD-ROM at the back of this text contains a folder, \data\uscitytm\. This folder contains historic surface temperature data for 172 U.S. locations. A file, Location.txt, lists the geographic locations along with their abbreviations used for the file names. Four data sets are provided per city: January low temperature, Janu-

ary high temperature, July low temperature, and July high temperature. Files are of variable size because of variable historic record. The file, Location.txt, also lists the time period over which data are recorded at each city.

a. Choose a city and use the program, *Visual_Data,* to determine using regression if a statistically significant temperature trend has occurred.

b. Which regression method is appropriate? Is error equally on X and Y, or only on Y? (Hint: X is time, and time is usually known precisely). These data sets contain values only for Y. X is implicit. Consequently, when using *Visual_Data,* respond when prompted that there is only one value per line in the data file. The program will then prompt you to enter the sampling interval for X. Respond by entering 1.0 for this sampling interval. For a more ambitious undertaking, you might analyze all of the data sets to determine all of the 172 locations, if any, that are associated with statistically significant temperature change. Notice that the phrase, temperature change, is used, and not global warming. Maybe some cities show a statistically significant cooling?

Judge statistical significance using the F-statistic computed for the slope of the linear regression. If the F-statistic is larger than about four, the regression is probably significant.

Multivariate Data Analysis

4

Based on results presented in Chapters 2 and 3, and in specific reference to the *Nevada_Landsat_6x* data set, we have concluded that none of the distributions of the seven variables is modeled well by a normal distribution. Moreover, bivariate combinations of these seven variables exhibit differing correlation, from excellent to only fair. These insights obtained through visualization are common for data collected and analyzed in the natural sciences.

In this chapter, we explore methods for visualizing relationships among more than two variables at once, rather than two at a time. We also explore methods for visualizing relationships among samples as a function of more than two variables. Collectively, this chapter is about *multivariate* relationships (also, *multisample* relationships). Because this text is about visualization techniques for understanding data, the primary focus of this chapter is on methods for creating pictures from data as an aid to understanding data relationships.

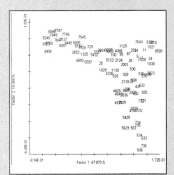

4.1 Analysis of Variance (ANOVA)

A method known as **analysis of variance** is a statistical hypothesis test for comparing the mean values of more than two variables. Abbreviated **ANOVA,** this test compares the null hypothesis:

H_0: Mean, variable 1 = Mean, variable 2 = $\cdots$ = Mean, variable M
against the alternative hypothesis
H_a: Mean, variable 1 $\neq$ Mean, variable 2 $\neq \cdots \neq$ Mean, variable M.

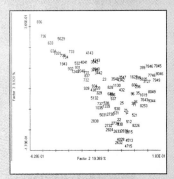

ANOVA is a quick way to determine similarity or dissimilarity among multivariate data. Typically, ANOVA is used by researchers that are able to perform controlled experiments in which manipulation of independent variables is possible.

ANOVA is valid if the following conditions are met:

1. a completely randomized design is used to collect the data; this requires the selection of any single variable in a process that is independent of the processes used to select the other variables;
2. the **populations** of the variables are represented well by the normal distribution model; a population is the collection of all possible values from which a subset of data is extracted; in the natural sciences, the entire population may be unknown or difficult to accurately define;
3. the variances of the populations must be equal for all variables.

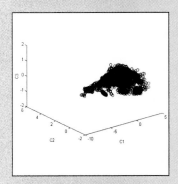

Using the *Nevada_Landsat_6x* data set as an example, suppose we wish to assess differences among the seven variables (seven Landsat TM bands)

using ANOVA. At once, we realize the first condition for applying ANOVA is violated by these data. The same instrument sampled the seven variables simultaneously for each particular pixel position. The same process, not different processes, was used to sample these seven variables. Second, the condition of normality certainly is not met by the collection of data, *Nevada_Landsat_6x*. But, the second condition for ANOVA is that the populations are represented well by the normal distribution model. Can we know the entire populations in this case? If we assume, because of the scanning method used by Landsat, that the collection of pixels representing a digital image is close to the actual population (pixels are continuous in space), then we conclude that we don't meet the second condition for ANOVA. Finally, are the population variances the same for the *Nevada_Landsat_6x* data? Probably not. Each variable represents a different portion of the electromagnetic spectrum. A particular frequency of electromagnetic energy is uniquely influenced by Earth's atmosphere and surface. Population variance may vary with frequency. We realize for the *Nevada_Landsat_6x* data that none of the conditions for a valid ANOVA are met.

This is most likely the conclusion for many data sets collected and analyzed in the natural sciences. Truly, one can argue the worthwhileness of ANOVA for the natural sciences. All this method will reveal is that there is a difference between at least two of the variables, but two (or more) that are different is not revealed. Nevertheless, maybe the conditions for a valid ANOVA can be relaxed for natural data and revealing insight can be gained, if for no other reason than to help quantitatively verify conclusions that are based on study of visualizations.

Returning to the example of the *Nevada_Landsat_6x* data, we saw in Chapter 2 that a normal distribution model does not represent well any of the seven data distributions. In Chapter 3, we saw that visible blue and visible green reflectances are highly correlated, and are linearly related. We also saw that the visible frequencies, blue, green, and red, are not as well correlated with near infrared, mid-infrared and thermal infrared frequencies. Is ANOVA sensitive to these visual observations?

ANOVA is actually an F-test. We learned about an F-test of statistical significance in Chapter 3 in application to regression analysis. Such a test compares variances. ANOVA is supposed to compare means, in fact the null and alternative hypothesis are so stated. In actual practice, ANOVA compares the variance of the mean values to the collective variance over all variables. The variance of the mean values, called **SST**, is computed as:

$$SST = n_1(\bar{x}_1 - \bar{x})^2 + n_2(\bar{x}_2 - \bar{x})^2 + \cdots + n_M(\bar{x}_M - \bar{x})^2$$

[the variance of the means equals weighted, squared differences between
the mean of a variable and the mean (average) of all *M* mean values; the
weights, *n*, are the number of values per variable]

In the equation for *SST*, $\bar{x}_j$ is the mean value for variable, *j*, and $\bar{x}$ is the mean value of the means for all *M* variables. In other words, if *M* is 3, and the means are 1, 2, and 3, then the average of these means is 2. Finally, n_j is the number of values representing variable, *j*. (In ANOVA, the *M* variables are not necessarily equally sampled; in fact, in a randomized design, the process used to collect one variable often yields a different number of data than the process used to obtain another variable. In the case of the *Nevada_Landsat_6x* data set, all variables are equally sampled).

The combined variances over all M variables, called **SSE,** is also a weighted sum:

$$SSE = (n_1 - 1)S_1^2 + (n_2 - 1)S_2^2 + \cdots + (n_M - 1)S_M^2$$

[the total variance, *SSE,* is equal to the sum of the *M* variances, each weighted by $n - 1$, where *n* is the number of data per variable]

Finally, the *F*-statistic is computed as the ratio, *MST/MSE*, for which $MST = SST/(M - 1)$ and $MSE = SSE/(N - M)$, and *N* is the total number of samples comprising the multivariate set. The computed value of *F* is compared to the table values for *F* using $M - 1$ numerator degrees of freedom and $N - M$ denominator degrees of freedom.

4.1.1   Application to the Nevada_Landsat_6x Data

Microsoft Excel is used for this demonstration. Middleton (2000) discusses the use of *MATLAB* for analysis of variance applied to regression analysis (as is shown in Chapter 3). The program, *Visual_Data,* does not provide a tool for ANOVA, except that which is included for the analysis of results from regression.

A user of *Excel* obtains an ANOVA as follows. Once the program is started, and the data file is opened in *Excel,* the user clicks Tools, then Data Analysis, then highlights the option, ANOVA, Single Factor and clicks Ok. The program then expects the user to define the Input Range. In this case, we desire an ANOVA over all seven variables, designated columns A through G by *Excel.* The Input Range is set to A$:G$ for all seven variables. Once this range is set, clicking Ok yields the ANOVA. Results for the *Nevada_Landsat_6x* data (from *Excel*): Anova: Single Factor SUMMARY

Groups	Count	Sum	Average	Variance
Column 1	10000	1485273	148.5273	592.7557303
Column 2	10000	734935	73.4935	228.9802558
Column 3	10000	1019478	101.9478	640.938769
Column 4	10000	864938	86.4938	416.1823798
Column 5	10000	1437893	143.7893	1348.307536
Column 6	10000	1974524	197.4524	328.1747517
Column 7	10000	876336	87.6336	621.939145

ANOVA

Source of Variation	MST	MSE	F	F crit
Between Groups	20062558.33	596.8	33616.9	2.10

In this example, the computed value for *F*, 33619.5, far exceeds the table value, *F*-crit, 2.1. Consequently, we reject the null hypothesis of equivalence among the seven means, and conclude that at least two mean values are significantly different. These results do match our visual impressions of these data. We expected this outcome.

4.2   Statistical Hypothesis Test for Two Data Sets

The foregoing application of ANOVA to the *Nevada_Landsat_6x* data suggests that at least two mean values of the seven are statistically significantly different. But, which two are different? Determining the answer to this question requires examining all possible bivariate combinations from the collection of seven variables. There are 21 possible combinations $(7 \times (7 - 1)/2)$, a daunting task. Reexamining the ANOVA, we note the differences in mean reflectance for these variables. Is there, for instance, a significant difference in the mean reflectances for visible blue and visible green, given that their mean values are 148.5 and 73.5 respectively, and their respective variances are 593 and 229?

A t-test can be applied for determining the statistical significance of the difference between these two mean values. The concept of the t-test is introduced in Chapter 3 for testing the significance of the correlation coefficient. In this case, we are comparing two sets of data, all visible blue and all visible green reflectances in the *Nevada_Landsat_6x* data set. The null hypothesis for this test is Mean, blue = Mean, green. The alternative hypothesis is that these two mean values are not equal.

This is a **two-tailed hypothesis test,** because the alternative hypothesis is stated simply as nonequivalence. We don't care if the mean for one is greater than or less than the mean for the other. The **tail** of a distribution is that portion associated with the smallest or largest values. The normal distribution, and many other distributions, have two tails, one for small values and one for large values. The *t*-statistic has a distribution known as the t-distribution. If the alternative hypothesis is proposed as nonequivalence, then we use both tails of the *t*-distribution for negative values (the mean for one is less than the mean for the other) and positive values (the mean for one is greater than the mean for the other). If the alternative hypothesis is stated as an inequality, less than or greater than, the type of hypothesis test is called a **one-tailed** test.

In either case, the *t*-statistic is computed as:

$$t = \frac{\text{Mean}_A - \text{Mean}_B}{\sqrt{\dfrac{S_P^2}{n_A} + \dfrac{S_P^2}{n_B}}}; \quad S_P^2 = \frac{(n_A - 1)S_A^2 + (n_B - 1)S_B^2}{n_A + n_B - 2}$$

[the t-statistic is equal to the difference between the two means, divided by the square root of the sum, pooled variance divided twice, once by the number of data in set, A, and again by the number of data in set, B. Pooled variance is equal to the sum of the two variances, S^2, each weighted by $n - 1$. This sum is divided by the total number of data in sets, A and B, less 2; this quantity represents the number of degrees of freedom for this test]

Applying this formula to blue and green reflectance from the *Nevada_Landsat_6x* data set gives the following results:

$$S_P^2 = \frac{(10000 - 1)593 + (10000 - 1)229}{10000 + 10000 - 2} = 411; \quad t = \frac{148.5 - 73.5}{\sqrt{\dfrac{411}{10000} + \dfrac{411}{10000}}} = \frac{75}{0.29} = 259$$

The critical value for *t*, two-tailed test, is approximately two. The computed value of 259 far exceeds this value. The null hypothesis is consequently rejected, and average blue reflectance is determined to significantly differ from average green reflectance. We have identified at least one of the bivariate pairs that contributed to the rejection of the null hypothesis in the ANOVA.

4.3   Principal Components Analysis

Hypothesis testing reveals a statistically significant difference in average blue and average green reflectance. This outcome is much different than what is visualized in Chapter 3 through correlation analysis. The correlation coefficient between visible blue and visible green reflectance for the *Nevada_Landsat_6x* data set is 0.98, very nearly perfect, positive correlation. What this shows is that two variables can differ substantially in magnitude, yet still be closely related.

How do the seven variables of the *Nevada_Landsat_6x* data set truly relate to one another as a function of their intercorrelation (a broader notion than bivariate correlation)? Furthermore, how do the samples in this data set relate to one another as a function of all seven variables? What is sought in this section are graphical methods for converting multivariate data to two dimensional plots to visually aid interpretation.

4.3.1   Principal Components

Graphical methods for displaying multivariate data relationships are based on using **principal components** to define coordinates for plotting variables and samples as points on two dimensional plots. The objective for developing these plots is a visual understanding of data associations. We start with a quantitative measure of association, such as correlation coefficient. From this basis, we derive auxiliary information, principal components, that are perfectly independent, or **orthogonal,** a necessary condition for use as orthogonal coordinate directions in a two dimensional plot.

Imagine a multivariate data set, such as the *Nevada_Landsat_6x* data set. Suppose we want to know how sample (pixel) 1 is related to sample (pixel) 1029, or sample 2525, or sample 3276. One approach is to graph all 10000 pixels choosing two of the seven variables to define X and Y coordinates. But, which two variables should be chosen? This decision will likely bias the analysis.

A more significant problem is that such a plot is a distortion if X and Y are intercorrelated (the correlation coefficient between X and Y is other than zero). The larger the absolute value of correlation coefficient, the more alike X and Y are and the greater is the distortion when plotting them as if they are independent. (Orthogonality is a visualization of zero correlation coefficient). Objective assessment of data through graphical methods mandates distortion-free plots. If two variables are going to be used to define orthogonal axial directions, the correlation coefficient for these two variables must be identically zero.

Principal components may be thought of as new variables extracted from the original group of M variables. If the M group of original variables is considered to represent an M-dimensional multivariate space, then a subset of P principal components may be thought of as representing a P-dimensional multivariate space in which there is no correlation between any two principal components. Graphically, the P-dimensional space may be visualized as a rotation, or transformation of the original M-dimensional space to yield the statistical independence among variables.

4.3.2   Standardized Principal Components Analysis

Correlation coefficient was briefly introduced in Chapter 2 and substantially defined in Chapter 3. In standardized principal components analysis, correlation coefficient is the metric (ruler) used to quantitatively describe data associations. The objective is to find a new set of variables, the principal components, that is an extraction of the original suite of M variables. In the case of the *Nevada_Landsat_6x* data, M is seven, the number of different spectral bands.

A matrix, $M \times M$ in size, is constructed. Each entry in this matrix is a correlation coefficient that describes the statistical relationship between two of the M variables. Perhaps the easiest way to view this matrix is as a data table. A particular row, or column of this table is associated with one of the M variables. Using the *Nevada_Landsat_6x* data set as an example, for which M is seven, the table is constructed as:

VARIABLE	1	2	3	4	5	6	7
1	1.	0.99	0.96	0.67	0.77	0.75	0.87
2	0.99	1.	0.98	0.73	0.81	0.74	0.89
3	0.96	0.98	1.	0.76	0.85	0.75	0.92
4	0.67	0.73	0.76	1.	0.84	0.54	0.76
5	0.77	0.81	0.85	0.84	1.	0.65	0.95
6	0.75	0.74	0.75	0.54	0.65	1.	0.73
7	0.87	0.89	0.92	0.76	0.95	0.73	1.

Notice that the correlation coefficient is equal to 1 (perfect, positive correlation) when the row and column numbers are equal. A row number represents one of the M variables, as does the column number. When the row and column numbers are equal, the correlation is that between a variable and itself. By logic, common sense, and definition, this correlation is perfect and positive. Notice as well that correlation coefficients for variables 1 (blue) and 2 (green), 1 and 3 (red), 1 and 4 (near infrared), 2 and 3, 2 and 4, 3 and 4, 3 and 6 (thermal), and 4 and 6 match those shown in Figure 3.12 for regression analysis applied to the *Nevada_Landsat_6x* data set. All of these correlation coefficients are positive, indicating like tendencies between bivariate pairs; when one is larger in value, likewise is the other, or one is smaller in value, likewise is the other. The poorest correlation is noted for the bivariate pair, near infrared (band 4) and thermal infrared. The highest level of correlation is noted between bivariate pairs composed of visible frequencies, blue, green, and red.

These values of correlation coefficient describe the 21 different bivariate relationships that are possible for a group of seven variables. Which method, however, was used to compute these correlation coefficients? For that matter, which method should be used? The supposition with much of the literature on principal components analysis is that the classical correlation coefficient, that which is derived from the slope found by applying linear, least squares regression in which all error is assumed on y, is used. This is not proper for the *Nevada_Landsat_6x* data set, a multivariate set of data for which the M variables are presumed associated with the same error imparted by sensor effects and natural interference. The better approach is to use the correlation coefficient derived from the assumption of equal error on x and y (Chapter 3 and York (1966)). These correlation coefficients are shown in the data table above.

4.3.2.1   Principal Components are Eigenvectors

Once the matrix of correlation coefficients is assembled, the next step involves decomposing this matrix into **eigenvalues** and **eigenvectors**. This is known as eigendecomposition, and qualitatively may be viewed as finding the "roots" of a matrix. The objective in this case is to extract information from the matrix of correlation coefficients. Components of this information are mutually orthogonal, which also represents statistical independence. By definition, eigenvectors are orthogonal.

Vectors represent spatial orientation. Consequently, eigenvectors represent orientation in P dimensional space, where P is a subset of M, the number of multiple variables comprising a set of data. If a matrix is eigendecomposable, there are as many eigenvectors as there are rows in the matrix. Because the ultimate objective in principal components analysis is the graphical display of data, and because eigenvectors are mutually orthogonal, eigenvectors are ideal for representing the axes (coordinate directions) when constructing two dimensional plots.

Each eigenvector is associated with an eigenvalue. Whereas eigenvectors represent orientation in M dimensional space, eigenvalues represent the level of importance assigned to each eigenvector direction when interpreting data associations. In standardized principal components analysis, M eigenvectors are extracted from the correlation matrix, and each is associated with an eigenvalue. These eigenvalues sum to M in standardized principal components analysis. That is, when eigendecomposing a matrix, the eigenvalues must sum to the trace of the original matrix. The trace is the sum of the diagonal entries. In a matrix of correlation coefficients, each diagonal entry equals one, and the sum of these diagonal entries is therefore M. The relative importance of each eigenvalue, expressed as a percentage, is equal to the eigenvalue divided by M (multiplied by 100 to convert to percent).

4.3.2.2   The Matrix Algebra

Now that the premise of principal components is established, the matrix algebra is briefly stated. Let a matrix to be eigendecomposed be called $[A]$. Let eigenvalues be represented by the symbol, λ, the most commonly used symbol representing eigenvalues in literature, and let eigenvectors be represented by $\{x\}$. The formal definition of eigendecomposition is as follows:

If an eigenvalue/eigenvector combination, λ and $\{x\}$, can be found such that $[[A] - \lambda[I]]\{x\} = \{0\}$, in which $[I]$ is an identity matrix of the same size as $[A]$, then the matrix, $[A]$, is eigendecomposable.

Not all matrices are eigendecomposable. Moreover, recall that eigendecomposition may be thought of as finding the roots of a matrix. This is the reason that the lefthand side must equal a matrix, $\{0\}$, whose entries all equal identically zero.

DEMONSTRATION

The 7×7 matrix of correlation coefficients for the *Nevada_Landsat_6x* data set is not easily used for a hand calculation demonstration of eigendecomposition. Instead, a very simple, 2×2, matrix is used for demonstration of the matrix algebra. In other words, let $M = 2$, and further assume the correlation coefficient between variables, 1 and 2, is 0.6. Then, the 2×2 matrix of correlation coefficients is

$$[A] = \begin{bmatrix} 1.0 & 0.6 \\ 0.6 & 1.0 \end{bmatrix}$$

We now write the formal approach to eigendecomposition: $[[A] - \lambda[I]]\{x\} = \{0\}$. This is now rewritten in expanded form:

$$\left[\begin{bmatrix} 1.0 & 0.6 \\ 0.6 & 1.0 \end{bmatrix} - \begin{bmatrix} \lambda & 0 \\ 0 & \lambda \end{bmatrix} \right] \begin{Bmatrix} x_1 \\ x_2 \end{Bmatrix} = \begin{Bmatrix} 0 \\ 0 \end{Bmatrix}$$

and regrouping terms gives:

$$\begin{bmatrix} (1.0 - \lambda) & 0.6 \\ 0.6 & (1.0 - \lambda) \end{bmatrix} \begin{Bmatrix} x_1 \\ x_2 \end{Bmatrix} = \begin{Bmatrix} 0 \\ 0 \end{Bmatrix}$$

If the matrix, $[A]$, is eigendecomposable, then positive, real roots can be found for the determinant of the composite matrix, $[[A] - \lambda[I]]$. In this example, the determinant, det, is equal to $(1.0 - \lambda)(1.0 - \lambda) - (0.6)(0.6)$ which, when expanded is equal to a quadratic expression: det = $\lambda^2 - 2\lambda + 0.64$. The roots of a quadratic expression are found as

$$\text{Roots} = \frac{-b \pm \sqrt{b^2 - 4ac}}{2a}$$

If real roots can be found, then the number of roots matches the order of the polynomial. The order of a quadratic expression is 2. In general for standardized principal components analysis, the order of the polynomial is equal to M. Furthermore, the eigenvalues are the roots of the determinant of the correlation coefficient matrix. In the formula for determining the roots of a quadratic expression, the coefficient, a, is the multiplier on the second order term; the coefficient, b, is the multiplier on the first order term, and c is the value of the constant. In this example, $a = 1$, $b = -2$, and $c = 0.64$. Consequently, the two roots are found as:

$$\lambda_1 = \frac{2 + \sqrt{4 - (4)(1)(0.64)}}{2} = 1.6; \quad \lambda_2 = \frac{2 - \sqrt{4 - (4)(1)(0.64)}}{2} = 0.4; \quad \lambda_1 + \lambda_2 = 2.$$

Because real-valued, positive roots were found, the matrix, $[A]$, is eigendecomposable. Further notice that the first eigenvalue is equal to the sum of either row of the correlation matrix, and the second eigenvalue is the difference between this sum and the trace of the matrix, 2.

Now that eigenvalues are determined, the challenge is to find the two eigenvectors. Starting with the first eigenvalue, 1.6, and recalling the expanded matrix system, we now can write:

$$\begin{bmatrix} (1.0 - \lambda) & 0.6 \\ 0.6 & (1.0 - \lambda) \end{bmatrix} \begin{Bmatrix} x_1 \\ x_2 \end{Bmatrix} = \begin{Bmatrix} 0 \\ 0 \end{Bmatrix} ; \text{using} \lambda_1 : \begin{bmatrix} -0.6 & 0.6 \\ 0.6 & -0.6 \end{bmatrix} \begin{Bmatrix} x_1 \\ x_2 \end{Bmatrix} = \begin{Bmatrix} 0 \\ 0 \end{Bmatrix}$$

From this, we can write two equations to solve for the eigenvector entries, x_1 and x_2:

$$-0.6\, x_1 + 0.6\, x_2 = 0; \text{ and } 0.6\, x_1 - 0.6\, x_2 = 0.$$

But, this is an indeterminate system. The reason for this is simple to explain. If an eigenvector, $\{x\}$, yields a product of $\{0\}$, so does a vector, $k\{x\}$, where k is any scalar value. Because k can be any real number, there are literally an infinite number of possible solutions for $\{x\}$! We instead solve for one of these possible eigenvectors by setting $x_2 = 1$, and solving for x_1 proportionately. In this case, the eigenvector is $\{1\ 1\}$. Using the second eigenvalue, 0.4, we find its eigenvector to be $\{-1\ 1\}$. Testing for orthogonality of the two eigenvectors, $\{1\ 1\}\ \{-1\ 1\}^T$ gives $(1)(-1) + (1)(1) = 0$. Two vectors are orthogonal if their cross product is equal to zero. Orthogonality is therefore demonstrated in this example.

Results of this eigendecompostion can be viewed graphically. The 2 rows of the original matrix, $[A]$, represent points in two dimensional coordinate space. The eigenvectors represent orientations in this original coordinate space in which the two original variables are statistically independent. A better plot, a distortion-free plot, for representing these two variables is one constructed using the eigenvectors as axes. Plots are shown:

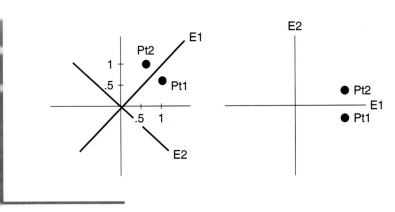

4.3.2.3   Eigendecomposing the Correlation Matrix for the *Nevada_Landsat_6x* Data

Visual_Data performs standardized principal components analysis. Correlation coefficients are computed on the assumption of equal error on x and y (Type II regression, Chapter 3). This is consistent with correlation coefficients shown earlier in the table for the *Nevada_Landsat_6x* data. Applying *Visual_Data* to these data for standardized principal components analysis yields the following seven eigenvalues:

Eigenvalue 1:	5.87	% of original correlation =	83.8
Eigenvalue 2:	0.54	% of original correlation =	7.7
Eigenvalue 3:	0.33	% of original correlation =	4.8
Eigenvalue 4:	0.21	% of original correlation =	3.0
Eigenvalue 5:	0.02	% of original correlation =	0.3
Eigenvalue 6:	0.02	% of original correlation =	0.3
Eigenvalue 7:	0.01	% of original correlation =	0.1
Total	7.00		100.0

Notice that the first eigenvalue, 5.87, represents about 84% of the original data information, correlation in this instance. These seven electromagnetic bands are highly intercorrelated. The stronger the amount of intercorrelation, the larger the first eigenvalue will be, moreover the lesser will be the values of the remaining eigenvalues. Notice further that the eigenvalues are listed in descending order. The first eigenvalue is always the largest and represents the most amount of original data information. The second eigenvalue is the second largest, and so on.

The ultimate objective of principal components analysis is graphical by projecting the original data onto the eigenvectors, then using two of the eigenvectors, one representing X and the other Y, when developing a two-dimensional plot. The projection is fairly simple. If the eigenvectors are loaded as columns in an $M \times M$ matrix, $[V]$, and the original data are represented by an $N \times M$ matrix, $[Y]$, then projecting the samples onto the eigenvectors is achieved by computing the $N \times M$ product, $[C] = [Y][V]$. The matrix, $[C]$, represents new plotting coordinates for the original samples. In this case, we have derived a **Q-mode** analysis, one that examines relationships only among the original samples.

Should we instead wish to focus solely on relationships among the M original variables, an **R-mode** principal components analysis is used. Of course, an ideal analysis is one that yields a simultaneous R-mode and Q-mode analysis. This enables an assessment of sample-to-sample, variable-to-variable, and sample-to-variable relationships, something that is not possible with either R-mode or Q-mode analysis alone. In the case of simultaneous R-mode and Q-mode analysis, new coordinates for the variables are obtained by

$$R\text{-mode Coordinate}_{jk} = \frac{-V_{jk}\sqrt{\lambda_k}}{\sqrt{\sum_{i=1}^{N} Y_{ij}}}$$

in which λ_k represents the kth eigenvalue. Once R-mode coordinates for the jth variable in the kth principal component direction are determined, Q-mode coordinates are calculated using the R-mode coordinates as:

$$Q\text{-mode Coordinate}_{ik} = \sum_{j=1}^{M} \frac{Y_{ij}\,R\text{-Mode}_{jk}}{W_i\sqrt{\lambda_k}}; \quad \text{for which } W_i = \sum_{j=1}^{M} Y_{ij}$$

R-mode coordinates are used to plot the new positions of the original M variables, and Q-mode coordinates are used to plot the new positions of the original N samples, all on one plot. This enables a simultaneous R-mode and Q-mode analysis. Two example plots are shown for the *Nevada_Landsat_6x* data. Figure 4.1 shows a two-dimensional plot based on the first two principal components, whereas Figure 4.2 shows a two dimensional plot based on the second and third principal components.

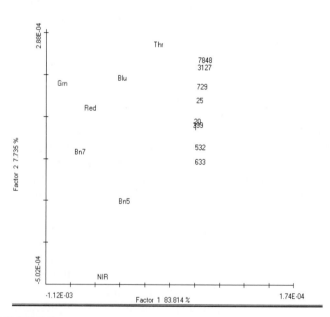

▲ FIGURE 4.1
Standardized principal components analysis of the *Nevada_Landsat_6x* data set, plot developed using the first two principal components. Numbers represent samples. The seven original variables are represented by plotting identifiers Blu, Grn, Red, NIR, Bn5, Thr, and Bn7. Sample 633, bottommost, is agricultural vegetation; sample 7848, topmost, is water. PC 1, Factor 1 on the graph, does not separate samples, but PC 2, Factor 2, does.

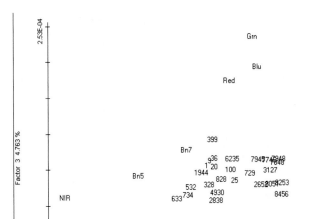

▲ FIGURE 4.2
Standardized principal components analysis of the *Nevada_Landsat_6x* data set, plot
developed using the second and third principal components. This graph is reviewed
in greater detail in the subsequent section on correspondence analysis.

Notice in these figures that the variables, Blu (Blue), Grn (Green), and
Red (Red) plot closely as a group, and NIR (near infrared, band 4), Bn5 (mid-
infrared, band 5), and Bn7 (mid-infrared, band 7) plot relatively closely as a
group, and Thr (thermal) is isolated by itself. This outcome is to be expected.
The visible portions of the electromagnetic spectrum are more closely related.
The near and mid-infrared frequencies are more alike. And, this outcome
shows that the thermal frequency is quite different. Landsat TM acquires images at about 10 A.M. to 10:30 A.M., local time. Solar, daytime heating has com-
menced and the thermal band largely shows variation in thermal inertia—how
quickly ground features warm. This attribute may not be closely related to visible or higher frequency infrared frequencies.

Samples, represented by numerical identifiers, plot as a group. The reason so few sample numbers are shown is attributable to a large amount of
overplotting, two or more samples occupying the same location in the graph.
Only one identifier can be printed. The program, *Visual_Data*, creates output
information showing which samples print over others. There is some resolution among the samples, and this is explored in later chapters, but samples
mostly cluster together in these plots.

4.3.3 Correspondence Analysis

There are many principal components methods from which to choose for
data analysis: principal components analysis, standardized principal components analysis, factor analysis, correspondence analysis, and several other
methods. All rely on eigendecomposition of some data similarity matrix
with the ultimate objective of developing two-dimensional plots. These various methods are distinguished with respect to the data similarity matrix. A
simple list is presented to describe this matrix:

1. Principal components analysis: similarity is based on variance and covariance;

2. Standardized components analysis: similarity is based on correlation coefficients;

3. Factor analysis: similarity is based on the cosine distance; or correlation coefficient; this method is unique in that it can rotate principal components;

4. Correspondence analysis: similarity is based on the chi-square distance between each data entry and its expected value.

In the preceding sections, we have explored the use of correlation coefficient for measuring the similarity among data. But, there are several algorithms for computing correlation coefficient, and which correlation coefficient to use—that which is based on error only on y, or on equal error on x and y, or that which is determined using least absolute deviation, or the one yielded by M-regression—affects the eventual outcome of the eigendecompositon.

There is the issue of robustness, a concept of insensitivity to data distribution. Do the data composing the multivariate set need to be normally distributed? Probably not, but if correlation coefficient is used and furthermore valid, shouldn't the residuals from the regression from which the correlation coefficients were determined be normally distributed? Technically yes, but in practice, how much can we get away with when it comes to these assumptions? Multivariate analysis presents a greater challenge when evaluating the correctness of assumptions. There is more than just the distribution of any one variable that is involved. There is also the notion of intervariable distribution, the distribution function describing the probability of one variable with another, so called **bivariate** probability density. Should this be a consideration when considering the principal components algorithm to apply to a multivariate data set? This deserves consideration, but there are no clear, or simple methods for testing bivariate distribution functions (or multivariate distribution functions when more than two variables are involved).

Robustness, when considered for principal components analysis, can be taken to mean consistency, or even stability, of analysis. In this sense, a method that is quite consistent and stable is **correspondence analysis.** This method examines the correspondence between rows and columns of a data matrix. Restated, correspondence analysis offers a simultaneous R-mode and Q-mode analysis to examine relationships between samples (rows) and variables (columns). In practice, this method is more robust than other principal components methods, a conclusion reached through much personal experience, as well as by other researchers (e.g., Jackson, 1993).

Robustness may be a concern with the *Nevada_Landsat* data set. We learned in Chapter 2 that none of the seven variables is represented well by a normal distribution model. What is more important to multivariate analysis is between variable relationships. We learned in Chapter 3, and earlier in this chapter, that some bivariate pairs from the group of seven variables are highly correlated, but other bivariate pairs are less well correlated. Correlation, however, is but one measure of closeness, or data similarity. Other similarity measures include variance and covariance.

Correspondence analysis uniquely applies a chi-square measure of closeness to describe data similarity. This measure is a function of the difference between a data value and its expected value, the same notion that was presented in Chapter 2 when testing data distribution models:

$$X^2 = \frac{(O - E)^2}{E}$$

[Chi-square is equal to the squared difference, observed minus expected, divided by expected]

In this formula, O represents an observed, or actual value, and E represents the expected value of O. Expected value is a probabilistic notion and is that value having the highest probability of occurrence.

In correspondence analysis, we start with our data, arranged in an $N \times M$ matrix, $[Y]$. All $N \times M$ entries in this matrix are then summed to yield a scalar quantity, **TOTSUM**. All entries in $[Y]$ are then divided by TOTSUM to yield a rescaled matrix, $[Y'] = [Y]/\text{TOTSUM}$. Once this step is complete, two vectors, $\{W\}$ and $\{T\}$, are formed from $[Y']$. The ith entry, $W(i)$, is the sum of the ith row of $[Y']$. The jth entry, $T(j)$, is the sum of the jth column of $[Y']$. There are N total entries in the vector, $\{W\}$, and M total entries in the vector, $\{T\}$. Given these definitions, the expected value of the matrix, $[Y']$, at the jth position on the ith row is equal to the product, $W\{i\}T\{j\}$.

Based on this information, a square, $M \times M$ matrix, $[S]$, is formed that will be eigendecomposed. Using the chi-square metric as a guide, this matrix is found as:

$$S_{kj} = \sum_{i=1}^{N} \left(\frac{Y'_{ij} - W_i T_j}{\sqrt{W_i T_j}} \right) \left(\frac{Y'_{ik} - W_i T_k}{\sqrt{W_i T_k}} \right), \quad j = 1, 2, \ldots, M; \quad k = 1, 2, \ldots, M$$

Suppose that $i = j = k = 1$. Then,

$$S_{11, \text{ for contribution 1 of } N} = \frac{(Y'_{11} - W_1 T_1)^2}{W_1 T_1}$$

Notice the analogy to the chi-square measure presented earlier. This helps demonstrate that the matrix, $[S]$, is composed of chi-square distances to represent data association, or closeness.

This matrix is decomposed into eigenvalues, placed on the diagonal of a matrix, $[D]$, and eigenvectors, arranged as columns of a matrix, $[V]$. From this information and the original, scaled matrix, $[Y']$, R-mode and Q-mode coordinates are computed as:

$$R\text{-mode Coordinate}_{jk} = \frac{-V_{jk} \sqrt{D_{kk}}}{\sqrt{T_j}}$$

in which j represents one of the original M variables and k represents a principal component. The R-mode coordinates are used to compute the Q-mode coordinates as:

$$Q\text{-mode Coordinate}_{ik} = \sum_{j=1}^{M} \frac{Y'_{ij} R\text{-Mode}_{ij}}{W_i \sqrt{D_{kk}}}$$

In this instance, i represents one of the original samples (rows of the original data matrix), and k represents one of the principal components.

4.3.3.1   Application to the *Nevada_Landsat* Data Set

Correspondence analysis is demonstrated by comparing its plots to those presented earlier for standardized principal components analysis. First, the eigenvalue summary is presented to show the relative importance of each principal component:

Eigenvalue:	1.079E-02	Percent Variation:	67.870
Eigenvalue:	3.079E-03	Percent Variation:	19.369
Eigenvalue:	1.578E-03	Percent Variation:	9.929
Eigenvalue:	2.489E-04	Percent Variation:	1.566
Eigenvalue:	1.764E-04	Percent Variation:	1.110
Eigenvalue:	2.480E-05	Percent Variation:	0.156

There are only six eigenvalues shown, even though M is seven for this analysis. In correspondence analysis, the value of the first eigenvalue is always one. It is known as the **trivial solution** and is discarded along with its eigenvector because this combination yields no particular insight to data intercorrelation. The remaining $M - 1$ eigenvalues/eigenvectors do provide useful discrimination of data associations and are used for developing plots.

For instance, a plot based on the first two eigenvectors represents 67.9 + 19.4, or 87.3% of the original data information. This plot is shown in Figure 4.3. Much better resolution is noted among the samples compared to the analysis obtained using standardized principal components. Labels for the seven variables are not shown in this figure. Each is plotted over by a sample as indicated in the following table:

Sample	293	prints over variable	Blu
Sample	34	prints over variable	Grn
Sample	11	prints over variable	Red
Sample	1036	prints over variable	NIR
Sample	24	prints over variable	Bn5
Sample	2001	prints over variable	Thr
Sample	1021	prints over variable	Bn7

Comparing this table to Figure 4.3 finds all seven variables toward the upper, right corner of the plot. Visible blue and green frequencies plot together, with visible red plotting between blue/green and the mid-infrared frequencies, Bn5 and Bn7. Near infrared plots a little away from the visible and mid-infrared frequencies, as does thermal infrared. These relationships noted among the spectral frequencies are similar, or identical to the relationships identified using standardized principal components analysis.

The most significant distinction between correspondence analysis and standardized principal components analysis for this example is the resolution obtained among samples (pixels). The following observations are made about samples: Sample 836 defines one extreme point of the graph at the lower right. Because it assigns numbers to each row of the data, *Microsoft Excel* is used to locate samples by number as shown in these plots. In Figure 4.3, we see that samples 836, 8456, and 8599 are associated with extreme points of the graph. These three samples have the following attributes:

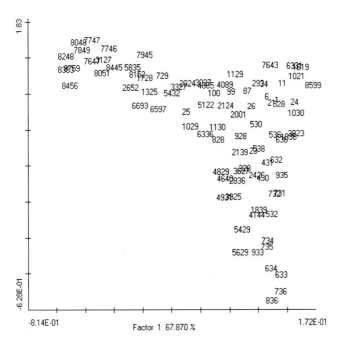

▲ FIGURE 4.3
Correspondence analysis of the Landsat Data Set (*Nevada_Landsat_6x_PC*). This plot
is developed using the first two eigenvectors.

Sample	Blue	Green	Red	NIR	Bn5	Thr	Bn7	Inference
836	86	36	31	196	90	156	25	vegetation
8456	81	35	33	10	10	165	3	water
8599	164	92	155	126	211	188	132	soil or rock

Each is associated with an inference that is based on the pixel values in
each of the spectral bands. Some agriculture exists within the scene from
which the *Nevada_Landsat_6x* pixels were extracted. Healthy, active vegetation, especially species having larger leaf surface area, is a strong reflector
of near infrared. Sample 836 has a relatively high NIR response, relatively
low visible response, and lower thermal response. This spectral variation is
consistent with vegetation. Water, on the other hand, absorbs near and
mid-infrared frequencies. Sample 8456 is inferred to be water because it
has higher visible blue reflectance, lower green and red reflectances, and
very low near and mid-infrared frequency response. Sample 8599 has a relatively high response across the entire spectral region represented by these
seven variables. This spectral response is consistent with some soil or rock
types. A lighter colored material is inferred because the thermal response,
188, is not particularly high. Thermal response is more a measure of thermal inertia because daytime illumination was active for several hours at
the time the Landsat TM scene was acquired. Similar sample associations
are found in a plot of the second and third principal components from correspondence analysis (Figure 4.4).

   If Figure 4.2 is revisited, a plot of the second and third principal components from standardized principal components analysis and one in which a
better resolution among samples is realized, we observe that samples 633,
8456, and 399 tend to define extreme portions of the plot. Sample 633 overprints

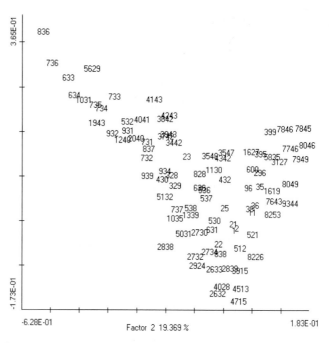

▲ FIGURE 4.4
A plot developed using the second and third principal components from correspondence analysis. A similar, 3-way association of samples is noted, in this case
defined by samples 4715, 7845, and 736. These samples are associated with samples (4715 = 8456 = water), (7845 = 399 = soil or rock), and (736 = 633 =
vegetation).

sample 836 and we infer that sample 633 represents vegetation. Sample
8456, of course, is inferred from correspondence analysis to represent water.
And, sample 399 overprints samples that plot very close to sample 8599 in
the correspondence analysis, so we infer that sample 399 represents soil or
rock. In summary, standardized principal components analysis and correspondence analysis tend to show the same relationships among the samples, and tend to separate samples into three categories: vegetation, soil,
and water.

4.3.3.2   Accommodating Missing Data in Correspondence Analysis

Suppose one or more values is missing in the original set of multivariate
data. Focusing on one of these missing values, further suppose that the jth
variable in the ith sample is missing. Because correspondence analysis is
based on a chi-square measure for data similarity, each entry in the scaled
matrix, Y', is associated with an expected value, $W(i)T(j)$. This concept of
expected value offers a unique way to replace, or estimate a missing value,
or values, using expected value. The algorithm is fairly simple and
straightforward:

1. Form the scaled matrix, $[Y']$, as described before; when computing
    the total sum of $[Y]$, use a value, zero, for a missing datum when
    computing this total sum; form the vectors, $\{W\}$ and $\{T\}$;
2. If a value is missing for $Y'(i,j)$, enter its expected value, $W(i)T(j)$,
    in its place;

3. Repeat step 1, but this time compute the total sum of $[Y']$, not $[Y]$; include the replaced estimates for missing data in this total sum; rescale the matrix, $[Y']$ using this new total sum; also compute new vectors, $\{W\}$ and $\{T\}$.

4. Because vectors, $\{W\}$ and $\{T\}$, are different after step 3, step 2 must be repeated; in other words, original missing data must be replaced again using the new, updated expected values, $W(i)T(J)$;

5. Steps 2 through 4 are repeated iteratively until the vectors, $\{W\}$ and $\{T\}$, no longer change.

6. Correspondence analysis then proceeds using the final version of the matrix, $[Y']$, and the final vectors, $\{W\}$ and $\{T\}$.

4.4 Multivariate, Linear Regression (Multiple Regression)

An additional aspect of multivariate data analysis is the extension of bivariate, linear regression to multivariate, linear regression of the form:

$$y^* = c_0 + c_1 x_1 + c_2 x_2 + \cdots + c_m x_m$$

[A dependent variable, y, is estimated based on a linear combination of m independent variables, x.]

in which the values, c, are coefficients (constants) multiplied to the independent variables. The collection of m independent variables, x, should ideally represent m statistically independent pieces of information, although in practice there is often some statistical interdependence among these variables. Should the correlation coefficient between any two independent variables, x, be close to 1 or to -1, one of the variables should be removed from the regression to avoid redundancy among the independent variables. All m variables, x, should be related in some way to y such that their inclusion in the regression aids estimation.

4.4.1 Logistic Regression: An Application of Multivariate, Linear Regression to the *Nevada_Landsat* Data Set

Throughout this text, the *Nevada_Landsat* data set is analyzed to show the relationship among the many different approaches to data analysis. In this case, the analysis of these data is obtained using **logistic regression.** This form of multivariate, linear regression predicts a value, y^*, having a numeric range, 0 to 1 inclusive. In other words, a logistic regression predicts the probability of something being present based on two, or more independent variables, x. In this application, we will use logistic regression to identify the presence/absence of vegetation. The data set, *Nevada_Landsat_clustered.dat* is used, wherein pixels representing different terrain features were selected based on an examination of the original image. Among the terrain features selected was agricultural vegetation, that which displayed the strongest reflectance in the near infrared. A new data set is developed, *Nevada_Landsat_clustered_logistic,* for which a dependent variable, y, is created by assigning $y = 1$ for all pixels identified as agricultural vegetation ($V1$), and all pixels not belonging to class, $V1$, assigned $y = 0$.

To minimize redundancy among the independent variables, an experiment is attempted to predict the logistic variable, y, using blue reflectance, near infrared reflectance, and thermal emission. Recalling the matrix of correlation coefficients for these data that was presented earlier in this chapter, the correlation coefficient between blue and near infrared reflectances is 0.67, between blue and thermal is 0.75, and between near infrared and thermal is 0.54. These are among the lowest correlation coefficients for any two of the seven variables, thus redundancy (a high degree of intervariable correlation) is minimized.

These three variables yield the following equation (from *Visual_Data*):

$$y^* = 0.64 + 0.00039 \, (\text{Blue}) + 0.0044 \, (\text{NIR}) - 0.0053 \, (\text{Thr})$$

which indicates that the larger the blue and near infrared reflectances, the larger will be the estimate of y, but the larger the value for the thermal emission, the lower the estimate of y will be. In other words, thermal emission is lower for healthy, dense vegetation. Anyone riding a bicycle next to an agricultural field in late evening will agree that the air is cooler over active crops.

This equation is used to classify the entire file, *Nevada_Landsat_6x.dat*. The goal is to test results determined through principal components analysis with respect to vegetation. *Microsoft Excel* is used to create a column of estimated values, y^*, using the foregoing equation. Resulting estimates are cited for a few examples:

Sample 633:	$y^* = 0.64$
Sample 736:	$y^* = 0.68$
Sample 399:	$y^* = 0.26$
Sample 7845:	$y^* = 0.01$
Sample 4715:	$y^* = -0.23$
Sample 8456:	$y^* = -0.16$

Samples 633 and 736 were inferred to represent vegetation in results yielded by correspondence analysis and standardized principal components. Multivariate, linear regression suggests a 64% to 68% probability that these samples belong to the vegetation class, $V1$. Moreover, multivariate, linear regression suggests that samples 399, 4715, 7845, and 8456 do not belong to vegetation class, $V1$. Indeed, these samples plotted far away from samples 633 and 736 in both correspondence analysis and standardized principal components analysis.

This example shows how several different methods of data analysis may be used in a complementary sense to help verify interpretations. To summarize applications to the *Nevada_Landsat* data set, three essential ground classes are identified: water, dense vegetation, and soil/rock/sparse vegetation. These three, uniquely different terrain types are identified based on all seven spectral variables associated with this set of data.

4.5   How Do I Reproduce Results in This Chapter, or Analyze My Own Data....

4.5.1   ...Using *Visual_Data*?

A data file is either created or opened in the upper window of *Visual_Data*. Any number of values may appear on each line. Once the file is opened (if creating a file from scratch, be sure to save it first using the Save option), click Tools, then choose the type of analysis that is desired:

1. If standardized principal components or correspondence analysis is chosen:

   a. *Visual_Data* will ask for the number of variables, M, used to describe each sample; M is seven for the *Nevada_Landsat_6x_PC data* (the principal components version of the *Nevada_Landsat_6x* data set);

   b. *Visual_Data* automatically counts the number of samples; an option is available for automatic assignment of plotting identifiers for the samples; if this option is chosen, numbers are assigned as plotting identifiers, starting from one; or, a user may specify plotting identifiers for samples by entering the identifier at the end of each data line; identifiers should be no longer than three characters and must be immediately preceded and followed by double quotation marks, such as "S1";

   c. If correspondence analysis is chosen, *Visual_Data* will ask if missing data are to be accommodated. If so, missing data must be represented using a consistent numerical value; because correspondence analysis cannot accommodate negative data values, a convenient numerical flag for missing data is −99 because a negative number is easy to spot when visually inspecting a data file;

   d. Finally, both tools ask a user to define the factor (principal component, eigenvector) to be used to define the x-axis direction, and that which is to be used to define the y-axis direction; the plot is then created on screen, and clicking Options provides an option for printing. Upon returning to the main program window, numerical results are found in the lower window. These include eigenvalues, followed by an overprint summary for the plot; this summary shows which samples in the plot overprint other samples or variables; or which variables overprint samples or other variables. The Save As option is used to save this output.

2. If multivariate, linear regression is the desired analysis option:

   a. any number of variables may appear on each line;

   b. *Visual_Data* will ask for this number of variables, then will ask how many of these will represent independent variables, and will ask a user to specify the position on each data line of each independent variable;

   c. Finally, a user will be asked which variable is to be treated as y. A plot of residuals then appears. Upon returning to the main program window, the coefficients for the regression are found in the lower program window. The Save As option is used to save these values.

4.5.2   . . .Using *Microsoft Excel*?

This chapter presents a unique relationship between *Visual_Data* and *Microsoft Excel*. *Visual_Data* lacks the capability to use coefficients obtained from multivariate, linear regression to classify different sets of data. Whereas *Visual_Data* can compute the coefficients of a multivariate, linear regression, some other program must be used to apply these coefficients to other data sets. A spreadsheet program, such as *Excel*, is ideally used for this application.

Results in this chapter were obtained using *Excel* as follows:

1. The data set, *Nevada_Landsat_6x_PC* was loaded into *Excel*; seven variables appear on each data line, and occupy columns A through G in *Excel*;

2. A function was then designed to fill column H as follows:

$$= A:A * 0.00039 + D:D * 0.0044 - F:F * 0.0053 + 0.64$$

3. The column, H, was then highlighted by clicking on the letter, H (this column should now be highlighted in black);

4. From the menu, Edit was clicked, the mouse pointer moved to Fill, then the word, Down, was clicked; this applied the foregoing function to every entry in column H. This function uses coefficients computed by *Visual_Data*. In this application, each sample in the original *Nevada_Landsat_6x* data set is associated with an estimated logistic variable that is an estimate of the probability that the pixel represents dense, healthy vegetation.

4.5.3   ...Using MATLAB 5.3?

Middleton (2000) presents MATLAB applications for both principal components analysis and multivariate, linear regression. To perform standardized principal components analysis on the *Nevada_Landsat_clustered_PC data,* the following MATLAB commands are used:

1. load the CD into the CD-ROM drive; suppose this drive letter is E; then switch to the directory that contains this data file:

```
>>cd E:\Nevada_Landsat_Data\
```

2. access the data file:

```
>> load Nevada_Landsat_clustered_PC_2.dat (this data
file is the same as Nevada_Landsat_clustered_PC, but
without the plotting identifiers for samples and
variables).
```

3. equate these data with a matrix, *X:*

```
>> X = Nevada_Landsat_clustered_PC_2
```

4. compute correlation coefficients:

```
>> R = corrcoef(X)
```

which yields the following results:

```
>> R = corrcoef(x)
R =
1.0000  0.9922  0.9733  0.4618  0.7412  0.9061  0.8952
0.9922  1.0000  0.9857  0.5253  0.7847  0.9058  0.9176
0.9733  0.9857  1.0000  0.5735  0.8437  0.9206  0.9521
0.4618  0.5253  0.5735  1.0000  0.8236  0.6146  0.6751
0.7412  0.7847  0.8437  0.8236  1.0000  0.8353  0.9487
0.9061  0.9058  0.9206  0.6146  0.8358  1.0000  0.9063
0.8952  0.9176  0.9521  0.6751  0.9487  0.9063  1.0000
```

These coefficients are different than what are presented earlier in this chapter. These particular correlation coefficients are determined by assuming all error is on *y* and that *x* is known precisely.

5. Eigendecomposition is performed using:

```
 0.6341   0.0621 -0.4647 -0.1866 -0.2420 -0.3707 0.3839
-0.6915   0.3542 -0.1025 -0.3225 -0.2125 -0.2880 0.3921
 0.1957 -0.1895  0.8210 -0.2257  0.0056 -0.2016 0.4002
 0.0158 -0.1464 -0.0364 -0.1874 -0.5089  0.7741 0.2892
 0.1693  0.6049  0.0150  0.0553  0.5736  0.3611 0.3778
-0.0880 -0.0118  0.0489  0.8773 -0.2455 -0.0924 0.3889
-0.2116 -0.6688 -0.3093 -0.0485  0.4979  0.0289 0.4014
```

Which yields the following (eigenvectas are shown first, followed by eigenvalues):

```
D =
  0.0032 0        0        0        0        0        0
  0        0.0066 0        0        0        0        0
  0        0        0.0122 0        0        0        0
  0        0        0        0.1087 0        0        0
  0        0        0        0        0.1402 0        0
  0        0        0        0        0        0.7700 0
  0        0        0        0        0        0        5.9592
```

Middleton (2000) developed a program, princomp, that executes all of these steps, plus plots the following: PC 1 versus PC 2; PC 2 versus PC 3; and a three-dimensional perspective developed by plotting PC1, 2, and 3 simultaneously. For example, a plot of PC1 versus PC 2 for the *Nevada_Landsat_clustered_PC_2* data is obtained by executing the following command:

```
>> [V,D] = princomp(X,0)
```

in which the 0 after the X represents standardized principal components; a plot of samples is then obtained:

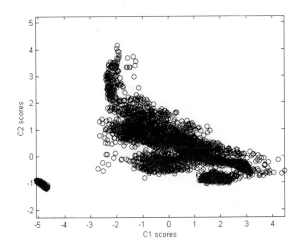

Notice that this plot shows the same pattern obtained using *Visual_Data*, a three-end point type of result, with samples associated with water plotting on the lower left, vegetation toward the top, and soil/rock/sparse vegetation tending to the lower right. A three dimensional perspective shows a similar pattern:

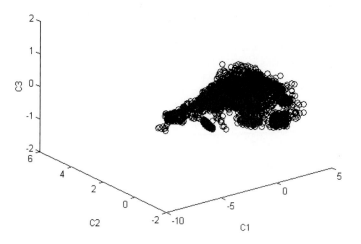

In these examples, we see how two programs, in this case *Visual_Data* and MATLAB, may be used to yield complementary analyses. This helps verify the correct execution of each program, moreover because they both yielded similar results for this particular data set, these complementary analyses help verify the interpretation that pixels in this Landsat scene define three principal ground types: water, dense vegetation, and soil/rock/sparse vegetation.

For multivariate, linear regression, Middleton (2000) developed a MATLAB program, mreg(X). First, a data file is loaded, in this example *Nevada_Landsat_clustered_logistic.dat*. The function, mreg, is written to expect only $p + 1$ values per data line, where p is the number of independent variables; y is expected to be the last entry on each data line. The data file, *Nevada_Landsat_clustered_logistic.dat*, has all seven Landsat spectral variables and the logistic value, y, is the eighth entry on each line. To replicate the analysis presented earlier in this chapter that is based on only three independent variables, blue, near infrared, and thermal, we first execute the following MATLAB commands:

1. Load the data:

   ```
   >> load Nevada_Landsat_clustered_logistic.dat
   ```

2. Equate the data with the array, X:

   ```
   >> X = Nevada_Landsat_clustered_logistic
   ```

3. Define a new array, Y, that will be compatible with mreg:

   ```
   >>Y(:,1) = X(:,1)        (Blue variable)
   >>Y(:,2) = X(:,4)        (Near infrared variable)
   >>Y(:,3) = X(:,6)        (Thermal variable)
   >>Y(:,4) = X(:,8)        (Logistic variable, y)
   ```

4. Then, apply Middleton's function, mreg:

   ```
   >>mreg(Y)
   ```

This yields the following result:

```
ans =
       0.0004
       0.0044
      -0.0054
```

a complementary result to that obtained using *Visual_Data*.

4.6   A Summary of Analyses Thus Far Obtained of the *Nevada_Landsat* Data Set

Through the first several chapters, we have observed that none of the seven spectral variables comprising the *Nevada_Landsat* data set have distributions modeled well by the normal distribution. The visible spectrum, blue, green, and red reflectance variables are highly intercorrelated, but are less well correlated with the near infrared, mid-infrared, and thermal infrared frequencies. In this chapter, principal components analytical methods show that pixels (samples) in this data set belong to three dominant groupings: water, dense, healthy (agricultural) vegetation, and soil/rock/sparse vegetation. From standardized principal components analysis, whereas the first principal component was not useful for separating samples, the second and third principal components were useful for separating samples associated with dense vegetation from those associated with water or higher reflecting, sparsely vegetated soil or rock. Correspondence analysis, because its first principal component, the trivial solution, is discarded, separated samples well with all of its principal components. Like standardized principal components, correspondence analysis showed a good separation between samples associated with dense vegetation and those associated with water or sparsely vegetated ground.

In the next chapter, we will learn how the pixels themselves relate to their closest, neighboring pixels as we explore the notion of spatial correlation.

4.7   Literature

Davis (1986) is an excellent reference for principal components analysis and correspondence analysis and was an inspiration for this chapter. Further in regard to correspondence analysis, Greenacre (1984) and Greenacre and Blasius (1994) are excellent sources for aiding in the understanding of this method, and especially for understanding the visual motivation behind principal components methods. As is the case throughout this text, Middleton (2000) is most helpful for understanding the application of MATLAB when performing data analysis, specifically multivariate data analysis in the case of this chapter.

4.8   Final Thoughts on the Robustness of Principal Components Methods

Both standardized principal components analysis (that which is predicated on the use of a correlation coefficient) and correspondence analysis (that which is based on a chi-square distance) yielded similar results in application to the *Nevada_Landsat* data set. Far from a rigorous mathematical proof, this is nonetheless empirical evidence of the robustness of these methods in application to data. Correspondence analysis often provides better resolution, but similar resolution can be achieved in standardized principal components analysis by not using the first principal component when constructing plots. Jackson (1993) arrived at a similar finding for benthic invertebrate data from 39 lakes in Ontario, Canada. In that study, principal components analysis based on the use of a correlation coefficient (standardized principal components analysis) and correspondence analysis provided the most consistent analytical results.

Exercises

1. Eigendecompose the following 2 × 2 matrices:

   a. $\begin{matrix} 1. & 0.2 \\ 0.2 & 1. \end{matrix}$ **b.** $\begin{matrix} a & c \\ c & b \end{matrix}$ **c.** $\begin{matrix} 9 & 4 \\ 4 & -3 \end{matrix}$

   d. $\begin{matrix} 0.5 & 4 \\ 4 & 0.4 \end{matrix}$ **e.** $\begin{matrix} 1 & 0 \\ 0 & 1 \end{matrix}$ **f.** $\begin{matrix} 1 & 1 \\ 1 & 1 \end{matrix}$

   g. $\begin{matrix} -2 & -1 \\ -1 & -2 \end{matrix}$ **h.** $\begin{matrix} 10 & 0.5 \\ 0.5 & 20 \end{matrix}$ **i.** $\begin{matrix} 1 & -0.5 \\ -0.5 & 1 \end{matrix}$

2. Compare standardized principal components analysis to correspondence analysis for the following data sets found on the CD in the directory, \data\othrdata\.

   a. *Stillwater Wildlife Refuge,* Churchill County, Nevada, 1985: two water sample data sets: StilJun.Dat (June, 1985) and StilSept.dat (September, 1985). Variables represent attributes of water chemistry (see Carr, 1995, Chapter 5). A hydrological application.

   b. Antelop.dat (distribution of antelope species in African wildlife refuges; see Greenacre, 1984); a biological application.

   c. Prob2A.dat through Prob2E.dat, all five data sets from Carr (1995, Chapter 5).

   d. Mineral1.dat, a rock geochemistry data set for understanding the usefulness of principal components analysis for economic geology;

3. Obtain a Landsat TM image and, using the Pixel Extraction Option in *Visual_Data* associated with the Digital Image Analysis Tool, extract a data set similar to the *Nevada_Landsat* data set and repeat some, or all, of the analysis presented in this chapter.

4. Apply logistic regression to the *Stillwater Wildlife Refuge* data found on the CD-ROM. Create a new logistic variable as follows: fresh water samples = 1; high TDS, high alkaline water = 0. Use the StilJun.dat data to compute the coefficients; the first two samples in this data set are "fresh" water; the others are not (Hint: see \data\othrdata\ and the file, stillgst.dat). Once the coefficients are computed, use *Excel* or some other spread sheet (even compute by hand) logistic values for each sample in the StilSept.dat data set. Which samples are found to be "fresh?" Use the following independent variables: boron (BO), arsenic (AS), specific conductance (SC). Or, try another combination of multiple variables.

Univariate Spatial Analysis

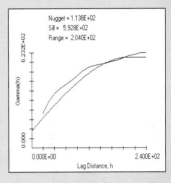

Thus far, we have examined data relationships solely on the basis of numerical value. Now, we will examine relationships as a function of both value and location in space. Space may be one-dimensional, including time to represent location, two-dimensional, often using coordinates, x and y, to represent location, three-dimensional, or some higher-order dimensional space. The key to this chapter is that additional information describing location (position) in space is included in analysis.

Because location in space is a key consideration, the term, *spatial analysis*, is introduced. An oft used synonym for spatial analysis is *geostatistics*. In this chapter, we will consider only a single variable that is distributed within space. Correlation will be the essential metric for understanding data relationships, but this metric will be a function of both value and location. The notion of spatial correlation is consequently introduced. Once spatial correlation is captured and understood, quantitative methods for spatial prediction are presented to yield visualizations of data in space.

5.1 Autocorrelation

Bivariate correlation analysis was described extensively in Chapter 3. Recall

$$m = Cov(x,y)/Var(x)$$

and

$$r = m(Sx/Sy)$$

therefore

$$r = Cov(x,y)/(SxSy)$$

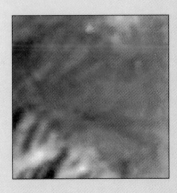

Suppose we consider correlation for only a single variable, Z, on the basis of two different observations, Z_j and Z_k. Then, we may write

$$r = Cov(Z_j, Z_k)/(Sz_j Sz_k) = Cov(Z_j, Z_k)/Var(Z)$$

In this instance, the term, $Cov(Z_j, Z_k)$, is known as the **auto-covariance** of Z. The prefix, auto, refers to self (singular), and auto-covariance represents covariance with self, or covariance between two different values of the same function. Auto-covariance is directly proportional to autocorrelation.

5.1.1 Introduction to Time Series Analysis

Let Z be a function of time, t. Then, the autocorrelation of Z may be written as

$$r = Cov(Z_{t1}, Z_{t2})/Var(Z)$$

A useful way of visualizing autocorrelation over time is by computing, then plotting a **power spectrum** of the function, $Z(t)$. Power represents the square of the Fourier transform of Z with respect to the spatial variable, t. This is called a transform because it transforms the spatial domain, in this case t, to the frequency domain, c. The discrete Fourier transform (that which can be calculated using a digital computer) is obtained as a convolution:

$$F(c) = \sum_{j=0}^{N-1} Z(t_j) * e^{2\pi icj}\delta t \approx \sum_{j=0}^{N-1} Z(t_j)\, e^{2\pi icj}\delta t$$

[The Fourier transform, F, is a measure of the degree of correlation between a function, Z, and a waveform having a wavelength, $1/c$]

That this is true is accepted more readily if one considers the Euler relationship:

$$e^{2\pi icj} = \cos(2\pi cj) + i\sin(2\pi cj)$$

in which i is the imaginary number $(\mathrm{SQRT}(-1))$.

In the Fourier transform, δt is the sampling interval for time, t (assumed regular). The function, $F(c)$, is the frequency domain analog of the spatial function, $Z(t)$. Because it represents a convolution (product) between the function, Z, and a waveform, the outcome, F, reveals cycles, if any, in Z. In this sense, F is a visualization of the autocorrelation of Z in time space, t.

DEMONSTRATION

In this example, a data file, ElNino.dat, is analyzed to determine if these data exhibit any significant periodicity (cyclic behavior). The program, *Visual_Data*, provides a tool for time series analysis of one-dimensional data strings. Opening the data file, Elnino.dat, then choosing this tool for an analysis yields the following power spectrum:

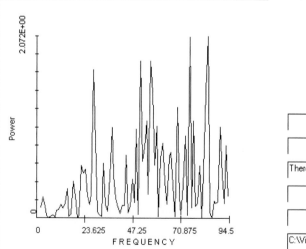

There are no statistically significant frequencie

C:\Visual_Data\other_data_sets\Elnino.dat

There are no statistically significant cycles present in these data, although the power for some frequencies appears to be sizeable. Statistical significance of cycles is judged relative to the cycle having the largest power, as is demonstrated next.

Stream runoff for Cave Creek, Kentucky (Haan, 1977) illustrates the identification of significant cycles in a power spectrum:

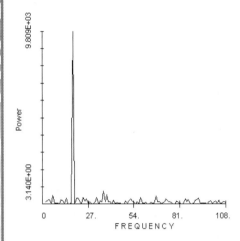

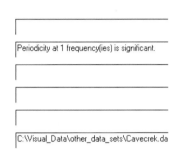

Periodicity at 1 frequency(ies) is significant.

C:\Visual_Data\other_data_sets\Cavecrek.da

In the case of the Cave Creek data, one significant cycle having a frequency of 18 is identified. The wavelength for this frequency, expressed as a length of time, is equal to 1/18 multiplied by the total time represented by these data. The total time, in months, is 216. The wavelength is consequently 216/18, or 12 months. In other words, there is a significant cycle present in these data exactly 12 months long. This suggests a seasonal influence on flow in Cave Creek. In fact, this creek is associated with a high discharge every March.

Judging the statistical significance of cycles is based on a technique borrowed from Davis (1986). A test statistic, W, is computed for each frequency, i, as $W_i = \text{Power}_i / (2\,Var(y))$, for which $Var(y)$ is the statistical variance of the original time series, y. Rather than comparing this test statistic to a table of values, a comparative value, g, is computed as:

$$g = 1 - e^{\left[\frac{\ln(\alpha) - \ln(N2)}{N2 - 1}\right]}; \alpha = \text{level of desired significance: 0.1; 0.05; 0.01.}$$

The program, *Visual_Data*, tests at the 0.05 level of significance. If the test statistic, W, is larger than the comparative value, g, the periodicity at frequency, i, is judged to be statistically significant.

If no frequency is found to represent a statistically significant periodicity, the power for any frequency may be purely coincidental. In this case, the autocorrelation between any two observations of the time series, y, is not explainable as a cyclic process. In contrast, the identification of at least one statistically significant period within the time series suggests a significant autocorrelation that is modeled well as a cyclic process. This is visualized for the Cave Creek data by setting all values, $F(c)$, equal to zero if c is not associated with a statistically significant power. Then, an inverse Fourier transform is performed as:

$$y(j) = \sum_{c=0}^{N-1} F(c) * e^{-2\pi icj}\delta c \approx \sum_{c=0}^{N-1} F(c)\, e^{-2\pi icj}\delta c$$

which yields the following filtered time series:

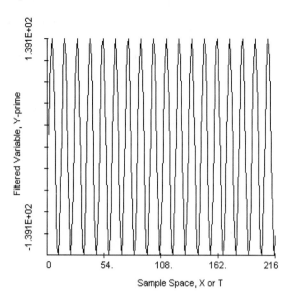

Comparing this filtered time series to the original time series helps to appreciate the distinctly cyclic nature of these data:

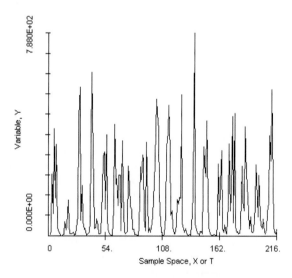

In the case of the Cave Creek, Kentucky data, the identification of a single, prominent cycle reveals a distinct periodic autocorrelation over space, T. Every 12 months, a peak is expected in stream discharge, with lesser discharge in between. This is quite useful information when predicting future levels of stream discharge. The El Nino data, in contrast, exhibited no distinct periodicity. The autocorrelation of these data, if any, is not a function of periodic behavior over the time interval of observation. A nonperiodic autocorrelation may be identified for these data and this is explored in the next section.

5.2   Spatial Autocorrelation

If not concerned with periodic behavior in data, but instead simply if a data value at one location is related to a value at another location, then a more general notion of spatial correlation, or spatial autocorrelation, is sought.

Why is information about spatial autocorrelation important? Examining spatial autocorrelation reveals something about the process, or processes, generating the data. With the Cave Creek, Kentucky, data, for instance, we see an annual cycle in these data, indicating that an annual (seasonal) process most significantly affects flow. In this case, runoff in the Spring contributes to the annual, high flow. With the El Nino data, however, we cannot find a simple explanation. Nonetheless, an analysis of autocorrelation reveals the very complex nature of these data and this is quite important to their understanding. An analysis of autocorrelation can also lead to the development of a model for predicting, or estimating, value at a location in space on the basis of autocorrelation.

5.2.1   The Variogram

An analytical tool for assessing spatial autocorrelation is the **variogram.** Perhaps a useful way to understand correlation is to substitute the term, **similarity.** A natural way to measure the similarity of two numbers is by computing their difference. This is the fundamental principle behind the type of analysis afforded by a variogram.

Recall the notion of statistical variance, a measure of the similarity of two numerical quantities, data value and mean. This statistic is a second order measure for several reasons, most significantly to render the difference between data value and mean a positive or zero quantity. The variance is computed as a sum over all data values. If raw difference is used, these differences will tend to cancel and the sum will be close or identical to zero. Squared differences sum to a positive, nonzero result.

The variogram is computed in the same way. A difference is the essential measure of similarity, and it is squared to obtain a positive or zero result. A summation is also used to obtain a cumulative assessment:

$$\gamma(h) = \frac{1}{2N} \sum_{i=1}^{N} [Z(x_i) - Z(x_i + h)]^2$$

> [The variogram, γ, is assessed for a distance, h, separating two data observations, $Z(x_i)$ and $Z(x_i + h)$, by computing their squared difference. All N possible pairs of data locations are found that are separated by more or less h, the squared difference computed for each pair, then all are summed. This sum is divided by $2N$ in the final step to yield the value of the variogram for a distance of separation, h (lag distance, h). In this sense, the variogram is one-half the average squared difference in data values separated by a distance more, or less equal to h.

This equation looks quite similar to the equation for statistical variance. In fact, the variogram is, like variance, referred to as a **second-order moment** for a set of spatial data. It should also be noted that, as written, this equation defines the **semi-variogram.** In practice, because it is always the semi-variogram that is computed, never the variogram (the variogram is $2\gamma(h)$), the prefix, semi, is discarded and hereafter only the term, variogram, is used.

Whereas it is a rather simple matter to write the formula for computing the variogram, it is an altogether tedious task to actually use this formula to develop the variogram by hand for a set of data. A simple exercise illustrates the extensive number of calculations involved.

DEMONSTRATION

The following spatial data configuration is used for this demonstration:

X:	1	2	3	4	5
Y					
5:	•65	•54	•64	•68	•109
4:	•70	•61	•76	•68	•90
3:	•82	•77	•83	•70	•76
2:	•77	•82	•79	•63	•58
1:	•95	•80	•57	•68	•85

When computing a variogram, we first must decide on the type of calculation. Do we simply want to determine spatial autocorrelation without worrying if it varies with direction? If so, we will use an *omnidirectional* calculation. On the other hand, if we suspect that spatial autocorrelation is anisotropic, or directionally variable, then we may use a directional calculation, restricting the calculation of a variogram only in a particular direction. This means that only data pairs separated by a distance, h, and oriented in a particular direction with respect to one another (e.g., east-west, north-south, and so on) are used. Once we decide on the type of calculation, we must "seed" the calculation by specifying what is known as the "class" size. This is the starting value, h, for the calculation, and is the incremental value for the calculation as h is gradually increased from the starting value to larger and larger values. The choice of class size is somewhat arbitrary and up to the analyst. But, class size should be relatively small in order to resolve spatial autocorrelation for smaller distances of separation. Class size might be chosen equal to the minimum sampling resolution for the data set. Or, class size may be chosen somewhat larger than this, yet still fairly small. For this demonstration, we will use a class size of one distance unit, the minimum sampling resolution in the data configuration above.

There is no way that a verbal discussion of this process can make clear how this calculation works. Instead, actual hand calculations are used for illustration to reinforce understanding. We begin with $h = 1$, the class size. An omnidirectional analysis is demonstrated. Pairs of data points are identified separated by a distance, W, such that $0 < W \leq 1$. Directional orientation is not a factor. Data pairs are listed using the following convention, $(a - b):c$, for which a and b are data values defining each pair, and c is equal to the square of the difference, $a - b$. We arbitrarily start at $X = 1, Y = 5$ to find data pairs:

(65 − 54):121	(54 − 64):100	(64 − 68):16
(68 − 109):1681	(70 − 61):81	(61 − 76):225
(76 − 68):64	(68 − 90):484	(82 − 77):25
(77 − 83):36	(83 − 70):169	(70 − 76):36
(77 − 82):25	(82 − 79):9	(79 − 63):256
(63 − 58):25	(95 − 80):225	(80 − 57):529
(57 − 68):121	(68 − 85):289	(65 − 70):25
(70 − 82):144	(82 − 77):25	(77 − 95):324
(54 − 61):49	(61 − 77):256	(77 − 82):25
(82 − 80):4	(64 − 76):144	(76 − 83):49
(83 − 79):16	(79 − 57):484	(68 − 68):0
(68 − 70):4	(70 − 63):49	(63 − 68):25
(109 − 90):361	(90 − 76):196	(76 − 58):324
(58 − 85):729		

Number of pairs, $N = 40$. Total sum of squared differences is 7750. $\gamma(1) = 7750/(2N) = 7750/80 = \underline{96.9}$.

The class size is also the amount that h is incremented by to investigate the spatial autocorrelation for larger and larger distances of separation (lag distances). In this demonstration, class size is equal to 1. Consequently, we increment h to $h = 1 + 1$, or $h = 2$. Now, we will identify all data pairs separated by a distance, W, such that $1 < W \le 2$. We *WILL NOT* include any data pairs found for the previous analysis, $h = 1$. We *WILL* include all pairs for which lag distance is equal to SQRT(2). Using the same convention for identifying pairs as was used in the previous calculation, we locate the following pairs (omnidirectional calculation):

(65 − 64):1	(54 − 68):196	(64 − 109):2025
(70 − 76):36	(61 − 68):49	(76 − 90):196
(82 − 83):1	(77 − 70):49	(83 − 76):49
(77 − 79):4	(82 − 63):361	(79 − 58):441
(95 − 57):1444	(80 − 68):144	(57 − 85):784
(65 − 82):289	(70 − 77):49	(82 − 95):169
(54 − 77):529	(61 − 82):441	(77 − 80):9
(64 − 83):361	(76 − 79):9	(83 − 57):676
(68 − 70):4	(68 − 63):25	(70 − 68):4
(109 − 76):1089	(90 − 58):1024	(76 − 85):81
(65 − 61):16	(54 − 76):484	(64 − 68):16
(68 − 90):484	(70 − 77):49	(61 − 83):484
(76 − 70):36	(68 − 76):64	(82 − 82):0
(77 − 79):4	(83 − 63):400	(70 − 58):144
(77 − 80):9	(82 − 57):625	(79 − 68):121
(63 − 85):484	(70 − 54):256	(61 − 64):9
(76 − 68):64	(68 − 109):1681	(82 − 61):441
(77 − 76):1	(83 − 68):225	(70 − 90):400
(77 − 77):0	(82 − 83):1	(79 − 70):81
(63 − 76):169	(95 − 82):169	(80 − 79):1
(57 − 63):36	(68 − 58):100	

Number of pairs, $N = 62$. Total sum of squared differences is 17593. $\gamma(2) = 17593/124 = \underline{141.9}$.

The number of computations involved in variogram analysis is considerable. This demonstration has thus far involved only two increments. Typically, variogram values are determined for up to 20 or more increments. The ultimate goal is a graph showing the variation of $\gamma(h)$ with h. Visual_Data is used to develop such a graph for the data configuration used in this demonstration:

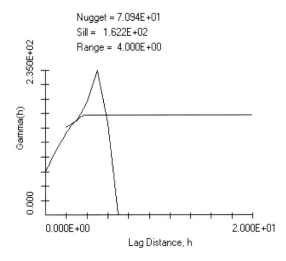

This particular data set is not spatially extensive enough to enable 20 increments of lag distance, h. Nevertheless, we are most interested in that portion of the variogram for smaller lag distance. In this example, variogram values increase as h increases, up to a limiting value of h, beyond which, at least for this data set, they decrease. The second line on this graph that shows a more ideal and smooth shape is an equation—a model—that is fitted to the computed variogram. Variogram models are discussed momentarily for spatial estimation.

As is emphasized in this demonstration, the goal of variogram computation is the development of a graph showing how the value of the variogram changes as lag distance increases. A spatial autocorrelation is inferred if the variogram increases with h in a more or less linear fashion. A parabolic and concave upward growth often is the result of a significant trend in the data. Typically, the variogram increases up to a limiting value of h, known as the **range,** at which it becomes more or less constant, or even decreases as in the demonstration presented earlier.

Beyond the range, there is no longer any spatial autocorrelation. Changes in h no longer influence the value of the variogram. This is an important aspect of variogram analysis. The range defines the spatial scale over which data are correlated. Smaller ranges result for data whose values change rapidly over space. Larger ranges result for data associated with more spatial regularity. Furthermore, the range is related to the scale of the spatial data.

The value of the variogram at the range is known as the **sill.** This is the more or less constant value that the variogram for spatial data often displays. Frequently, and theoretically, the sill value is equal to the statistical variance for the spatial data values. This is the reason that the variogram is computed by dividing the cumulative sum of squared differences by $2N$, rather than N. Otherwise, the sill value would equal twice the statistical variance. The advantage associated with a sill value more or less equaling statistical variance is discussed later in this chapter.

Quite often, if the graph of the variogram is extended back to the origin, $h = 0$, it will be found to intersect the vertical axis at a value, $\gamma(h) > 0$. This is a typical circumstance for many spatial data sets. A nonzero value of the variogram at the origin is known as the **nugget** value. Variogram analysis was developed for applications in the minerals industry (Matheron, 1963). A nugget value represents a degree of randomness, perhaps white noise, in the spatial data. Alternatively, the nugget represents micro spatial autocorrelation at a scale below the minimum sampling resolution. Regardless, imagine a placer gold deposit. We sample at a point, (x,y), and find a nugget. We sample just a fraction of a distance away, (x',y'), and find nothing. This illustrates micro spatial variation (or randomness, however you wish to consider this example).

DEMONSTRATION

Variogram analysis of the El Nino data. Previously, these data were found not to be associated with any statistically significant cycle. We further assess these data for average, spatial correlation in time vis-a-vis the variogram. The variogram shows what is known as a "nugget effect." The nugget and sill are equal; the variogram is more or less constant and flat over all lag distances. This result is consistent with a function showing no significant spatial autocorrelation:

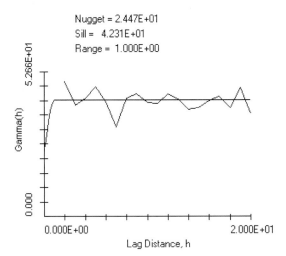

The attempted model fit to this variogram suggests that the nugget is less than the sill. *Visual_Data* attempts a model fit by setting the nugget value equal to one-half the value of the variogram for the first increment of lag distance. In this case, the variogram value for the first lag increment is about equal to the sill:

#Pairs	Gamma(h)	Average Distance, h
188	5.266E + 01	1.000E + 00
187	4.893E + 01	2.000E + 00
186	4.059E + 01	3.000E + 00
185	4.297E + 01	4.000E + 00
184	4.701E + 01	5.000E + 00
183	4.180E + 01	6.000E + 00
182	3.242E + 01	7.000E + 00
181	4.309E + 01	8.000E + 00
180	4.444E + 01	9.000E + 00

Fractal Analysis:
Following Results Pertain to 4-step intervals of the Semi-Variogram:

> Interval of Semi-variogram: Lag 2 To 5
> Slope, this interval $= -4.84456809771257E-02$
> Fractal Dimension, this interval $= 2.02422284048856$

Of further note is the assessed value of fractal dimension, approximately 2.0, for these data based on the first few increments of lag distance. This fractal dimension is consistent with white noise, or a completely random process over time. The fractal dimension, nugget effect variogram, and power spectrum outcomes all suggest a lack of spatial autocorrelation in time for these data.

5.2.1.1 Application to the *Nevada_Landsat* Data Set

A focus is on the visible blue variable in this section for the sake of brevity. The *Nevada_Landsat_6x* data set was obtained by sampling the original Landsat TM image at every sixth pixel of every sixth row. The minimum sampling resolution for this data set is therefore 6 (pixels) and this value is used as the class size. (It is acknowledged that Landsat TM pixels represent a ground resolution size of 30 m × 30 m. A 6 pixel minimum sampling resolution physically represents 180 m. But, row and pixel values are used as coordinates in the *Nevada_Landsat_6x* data set, and class size must be consistent with the way in which coordinates are defined.) The variogram for the visible blue variable is shown in Figure 5.1.

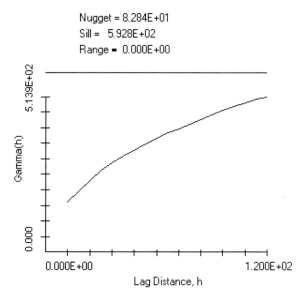

▶ FIGURE 5.1
Variogram for the visible blue variable, *Nevada_Landsat_6x.dat* data set.

This figure shows an increasing variogram with lag distance, *h*. But, 20 increments of lag distance were not sufficient to resolve the range. That is, the variogram is still increasing at the 20th lag increment. Figure 5.1 attempts to model the variogram by setting the sill exactly equal to the statistical variance of the data analyzed. In this case, the variogram has not attained this sill at the 20th lag increment. Another calculation is attempted, this time setting class size equal to 12, twice the minimum sampling resolution. This will enable an analysis of spatial autocorrelation up to a distance of 240 pixels at the 20th lag. This outcome is shown in Figure 5.2.

Even at a lag distance of 240 pixels, the variogram for the visible blue variable has not attained a sill. The range is inferred to be greater than a 240 pixel distance. Notice, though, that the variogram shows a bilinear character (Figure 5.2), with a break in slope that occurs at a lag distance of approximately 120 pixels. For lag distances between 0 and 120 pixels, the variogram increases more rapidly than for lag distances between 120 and 240 pixels. This is an example of **nested** spatial autocorrelation. The nature of the autocorrelation changes at particular lags, yet a range has not yet been reached. This is best modeled using a linear combination of two, or more valid variogram models.

Visual_Data, and many other programs, allow for nested variogram modeling. This is, however, a process that may require some experimentation to figure out the right combination of models to use, moreover their pa-

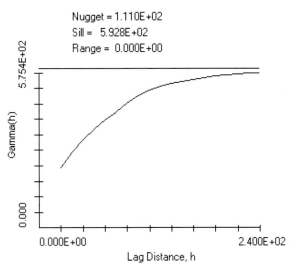

▲ FIGURE 5.2
Variogram of the visible blue variable, *Nevada_Landsat_6x.dat* data set, using a class size equal to 12. The variogram is still slightly increasing at a lag distance of 240 pixels and has not quite attained a sill value. The range, in this case, of the spatial autocorrelation exceeds 240 pixels.

rameters. In the case of the visible blue reflectance, the following two model combination is considered (see Section 5.4.1 on variogram models):

$$\gamma_1(h) = 111 + 400\left(\frac{1.5h}{120} - \frac{0.5h^3}{120^3}\right), h < 120; \quad \gamma_1(h) = 511, h \geq 120$$

$$\gamma_2(h) = 0 + 82\left(\frac{1.5h}{240} - \frac{0.5h^3}{240^3}\right), h < 240; \quad \gamma_2(h) = 82, h \geq 240$$

$$\gamma_{Total}(h) = \gamma_1(h) + \gamma_2(h)$$

Notice that all of the nugget is placed on the first model, and its sill is that value of the variogram where a noticeable change in the slope of the variogram occurs. Its range is that value, h, at which this change of slope occurs. The total range, 240, is put on the second model. Its sill is equal only to that small portion of the variogram between the first change in variogram slope and the ultimate sill value. A graph of the total model is shown for comparison to Figure 5.2:

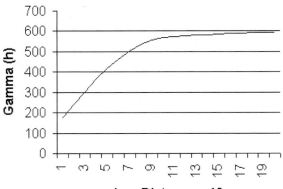

Notice that this nested combination of models closely resembles the spatial autocorrelation exhibited by the visible blue reflectance data.

When a satellite detects and records blue reflectance from space, a good deal of the blue reflectance is due to Earth's atmosphere. This atmospheric interference imparts a haze to a visible blue image. The nested nature of the visible blue variogram is inferred to be the result of the atmospheric interference. That portion of the variogram between lags 0 and 120 pixels is more greatly influenced by actual blue color variation on Earth's surface, whereas that portion of the variogram between lags 120 and 240 pixels is more greatly influenced by atmospheric haze. This is a possible explanation for the nested spatial autocorrelation.

5.2.1.2   Directional Variogram Analysis

Variogram analysis shown in Figures 5.1 and 5.2 represents an omnidirectional (all directions, or isotropic) analysis of the visible blue data for the *Nevada_Landsat_6x.dat* data set. Suppose we suspect that spatial autocorrelation changes with direction, where direction is defined on Earth's surface using the 360 degrees of a circle. For instance, 0 (or 180) represents east-west; 90 (or 270) represents north-south; 45 (or 225) represents northeast-southwest; 135 (or 315) represents northwest-southeast; and so on. In fact, we may analyze any particular compass direction, 0 to 360. Especially if our data are geologically controlled, or the result of prevailing wind direction, are the result of flow direction, or some other directionally dominant process, then we must consider that spatial autocorrelation for our data may exhibit an anisotropic, directionally variable character. Typically, directional variogram analysis yields four directional computations: east-west, northeast-southwest, north-south, and northwest-southeast. Because data may be only approximately oriented in one of these directions, a tolerance of ±22.5 degrees is allowed.

The goal of directional variogram analysis is to determine if the **range** of spatial autocorrelation resolved by each directional variogram is approximately the same (isotropic autocorrelation) or differs with direction (anisotropic autocorrelation). Differences in ranges should be evaluated against the level of noise in the data. Furthermore, the smaller the data set, the more influential is this noise in directional analysis. For instance, if ranges differ by no more than 20 percent for particularly noisy variogram outcomes, isotropic autocorrelation may be inferred. In the case of the Landsat Data, the directional outcomes shown in Figure 5.3 are not at all noisy. This data set consists of 10000 spatial samples, a rather large data set providing substantial information for directional analysis. More weight may be given to differences in ranges and the inference of anisotropic behavior.

5.3   Fractal Geometry

Variogram analysis defines the self-affine fractal dimension of an object or process. Self-affine geometric translation is one in which one spatial direction is changed differently from another to preserve the statistical relationship between these directions. Affine geometry describes objects or processes that scale differently with dimension. In a plot of the value of gold over time, for instance, gold value is not the same measure as time; dollars (or pounds; or pesos) are a different notion than seconds; minutes; days; or years. The geometry of x (time) scales differently than that of y (value). This represents an affine geometry.

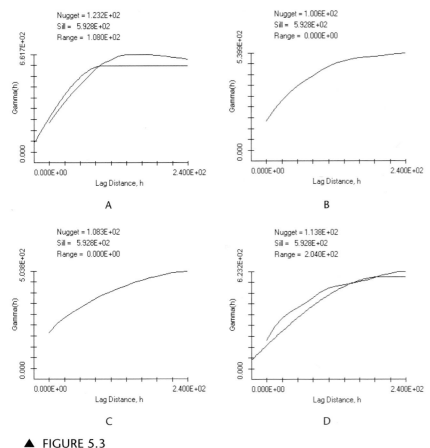

▲ FIGURE 5.3

Directional variograms for the visible blue variable, *Nevada_Landsat_6x.dat* data set. A. East-West; B. Northeast-Southwest; C. North-South; and D. Northwest-Southeast.

Suppose the daily price of gold is recorded for one year. Should this time series be resampled keeping only the price of gold every other day (price is resampled by a factor of 2), then we must rescale the price of gold by 2^H to preserve the statistical moments (specifically the variogram) of gold value. This is a self-affine rescaling relationship. The superscript, H, is known as the Hurst dimension (after Hurst, 1951). We can determine the Hurst dimension from spatial autocorrelation information vis-a-vis the variogram. The Hurst dimension, H, is equal to one-half the slope of the plot of the natural logarithm of variogram value versus the natural logarithm of lag distance (for smaller lag distances). Although the power spectrum can also be used to determine the Hurst dimension, the variogram yields a more straightforward and stable procedure for determining H (Carr and Benzer, 1991).

Consider the following expression for the variogram written theoretically as an expected value:

$$2\gamma(h) = E[Z(x_i) - Z(x_i + h)]^2 \propto Abs(h)$$

Notice that this reinforces what is stated earlier about the behavior of the variogram near the origin. This theoretical notion relates the expected value of squared deviations of data value to a **linear** variation in h, lag distance. It is further shown in many texts that discuss self-affine fractal processes (e.g., Carr, 1995) that the variograms for such processes obey the following law:

$$2\gamma(h) = E[Z(x_i) - Z(x_i + h)]^2 \propto (Abs(h))^{2H}$$

If the log of $\gamma(h)$ is plotted versus the log of h, the slope of the plot is theoretically equal to $2H$.

By determining H using the variogram, we know how to rescale the value of spatial data upon resampling to obtain the same variogram as that for fully sampled data.

DEMONSTRATION

Visual_Data is applied to Rainfall.dat, rainfall amounts recorded during a storm in Boston, Massachusetts 25 October 1980 (Rodriquez-Iturbe, et. al., 1989). When used to analyze one-dimensional data vis-a-vis the variogram tool, *Visual_Data* automatically computes *fractal dimension* for different increments of lag distance, smaller to larger. Fractal dimension, D, is related to Hurst dimension as $D = 2 - H$. The following results are presented for the rainfall data:

Fractal Analysis:
Following Results Pertain to 4-step intervals of the Semi-Variogram:

Interval of Semi-variogram: Lag 2 To 5
Slope, this interval = 0.702172571566964
Fractal Dimension, this interval = 1.64891371421652

Interval of Semi-variogram: Lag 6 To 9
Slope, this interval = 0.361011874047506
Fractal Dimension, this interval = 1.81949406297625

Interval of Semi-variogram: Lag 10 To 13
Slope, this interval = 0.435054973603608
Fractal Dimension, this interval = 1.7824725131982

Interval of Semi-variogram: Lag 14 To 17
Slope, this interval = 0.898590818599531
Fractal Dimension, this interval = 1.55070459070023

These results show that an assessment of fractal dimension varies depending on the lag interval used. That portion of the variogram closest to the origin (for smaller lag distances) is more appropriate, given the expected value relationship between variogram and Hurst dimension, H. In the case of these data, we therefore conclude that a fractal dimension of 1.65 is an accurate measure.

Fractal dimension may also be estimated using the power spectrum. A linear regression is applied using y, ln(power), and x, ln(frequency). Fractal dimension is equal to $(5 - Abs(m))/2$, where m is the slope of this regression. The type of regression to apply is that which assumes all error is on y (the natural logarithm of power). The power spectrum for these rainfall data yields a regression with slope, m, equal to -1.28, yielding a fractal dimension of 1.86, a greater estimate than that obtained from the variogram for shorter lag distances, but comparable to that obtained from the variogram for lag distances of 6 to 9.

If fractal dimension, D, is equal to $(5 - Abs(m))/2$, and D is equal to $2 - H$, then it can be shown that $m = -(2H + 1)$. Recalling m is the slope of the linear model relating ln(Power) to ln(frequency), then we can write ln(Power) = ln(Var(y)) $- (2H + 1)$ ln(frequency). This expression implies that the power for a frequency of zero is equal to the statistical variance of the time series. From the variogram we know that $\ln(\gamma) = 2H \ln(h)$. Solving each relationship for H yields a relationship between the power spectrum and the variogram:

$$\frac{\ln(Var(y)) - \ln(Power)}{\ln(frequency)} = \frac{\ln(\gamma)}{\ln(h)} + 1$$

This relationship holds for any frequency, f, related to lag distance as $h = N/f$, where N is the length of the one-dimensional series.

In concluding this section, the notion is forwarded that a spatial phenomenon exhibiting spatial autocorrelation that is proportional to lag distance, h, is fractal. The slope of $\ln(\gamma)$ versus $\ln(h)$ yields a line having a slope equal to $2H$, and H is the Hurst dimension. Fractal dimension is determined from the Hurst dimension as follows: A. for one-dimensional space, $D = 2 - H; 1 \leq D \leq 2$; B. for two-dimensional space, $D = 3 - H; 2 \leq D \leq 3$; C. for three-dimensional space, $D = 4 - H; 3 \leq D \leq 4$; and D. for any order dimension, $D = K - H$, where K is the dimension measure. Notice that fractal dimension is always intermediate between the dimension of analysis and the next higher dimension.

5.4   Kriging: Spatial Interpolation as a Function of Spatial Autocorrelation

Once the spatial autocorrelation of a spatially distributed variable is known, this information may be useful for predicting the value of the spatially distributed variable at an unsampled location. Matheron (1963) developed a Gaussian least squares estimator for spatial data that he called **kriging** in honor of Dr. D.G. Krige, a South African mining engineer who was the first to attempt ore reserve estimation using a probabilistic model. Kriging is a weighted average estimation algorithm of the form:

$$Z^*(x_0) = \sum_{i=1}^{N} \lambda_i Z(x_i)$$

[The estimated value is equal to the sum, over N closest spatial locations, of the products obtained by multiplying the observed or sampled value at each of the N spatial locations by the weight for that location; the weight will be a function of the autocorrelation between the neighboring location and the estimation location.]

Z represents what is known as a spatially distributed **random function.** $Z(x)$ is standard notation for the value of the random function at a spatial location, x. This notation is somewhat unfortunate because it represents a large potential for misunderstanding. The "x" in this notation is not an x-coordinate. Rather, x is a point in space that is defined by x, y, Z coordinates. The notation, x_0, represents the spatial location at which an estimate of Z is desired. The notation, Z^*, wherein an asterisk is used with Z, represents the estimated value. Moreover, $Z(x_i)$ are N spatial locations at which Z was observed or sampled in close proximity to the estimation location. These N locations are in fact the *closest* locations to the estimation location. Because these N values of Z are already known, the only unknowns in this equation are the N weights, λ, to be multiplied to the N values of Z to yield the estimated quantity, Z^*. Information that defines the spatial autocorrelation of Z is used to solve for these weights.

It has already been stated that kriging is a Gaussian, least squares estimation method. Moreover, it is a local (neighborhood defined by N) and a linear estimator. However, developing a formula for solving for the weights, λ, requires a probabilistic notion for squared error because we do not know the true value of Z at an estimation location. Recalling linear, least squares regression, Chapter 3, we use an analogy to develop a solution for the weights in kriging. Estimation error is written as an expected value: $E[Z(x_0) - Z^*(x_0)] = 0$. That is, the expected value of error, that error which has the highest probability of being true, is equal to zero. Using this expression for error, we further write that $E[Z(x_0) - Z^*(x_0)]^2 =$ error variance = minimum. In kriging, we see that the expected value of the square of the error is minimized, moreover that this expected value is equivalent to the variance of the estimation error.

This notion of error variance is now expanded:

$$\text{error variance} = E[Z(x_0) - Z^*(x_0)]^2$$

$$= E[Z(x_0)]^2 - 2E[Z(x_0)Z^*(x_0)] + E[Z^*(x_0)]^2$$

We now see that error variance is a function of three components, all written as an expected value. Expected value is a theoretical and probabilistic notion. Often, unless precise information is available for explicitly defining probability, we cannot directly solve for an expected value. Instead, we can search for a way to substitute explicitly defined information for the expected value.

For example, statistical variance is introduced in Chapter 2, there using a discrete (can be calculated) definition. Statistical variance also has a probabilistic definition based on expected value:

$$Var(Z) = E[Z]^2 - \overline{Z}^2; \rightarrow Var(Z) + \overline{Z}^2 = E[Z]^2$$

Comparing this relationship to Component 1 of the expression for error variance, we see that Component 1 can be rewritten as $E[Z(x_0)]^2 = Var[Z(x_0)] + Mean^2(Z)$.

The notion of covariance was introduced in Chapter 3. It, too, has a probabilistic definition based on expected value:

$$Cov(x,y) = E[xy] - \overline{x}\,\overline{y}$$

and for autocovariance (written for Z):

$$Cov(Z(x_i),Z(x_j)) = E[Z(x_i)Z(x_j)] - \overline{Z}^2;$$

$$\rightarrow E[Z(x_i)Z(x_j)] = Cov(Z(x_i),Z(x_j)) + \overline{Z}^2$$

This enables rewriting the second component of error variance as:

$$-2E[Z(x_0)Z^*(x_0)] = -2Cov(Z(x_0),Z^*(x_0)) - 2\overline{Z}^2$$

Recalling the formula for kriging, we may substitute for $Z^*(x_0)$ to obtain for this component

$$-2\left[Cov\left(Z(x_0)\sum_{i=1}^{N}\lambda_i Z(x_i)\right)\right] - 2\overline{Z}^2 = -2\left[\sum_{i=1}^{N}\lambda_i Cov(Z(x_0)Z(x_i))\right] - 2\overline{Z}^2$$

Finally, for the third component, we may start by substituting the formula for kriging for $Z^*(x_0)$:

$$E[Z^*(x_0)]^2 = E\left[\sum_{i=1}^{N} \lambda_i Z(x_i)\right]^2$$

It is shown in many sources (e.g., Carr, 1995) that this is equivalent to

$$\left[\sum_{i=1}^{N}\sum_{j=1}^{N} \lambda_i \lambda_j \, Cov(Z(x_i)Z(x_j))\right] + \bar{Z}^2$$

Expected value is now replaced in all three components by information defining variance and covariance for the spatially distributed variable, Z. We may now rewrite the expression for error variance. Calling the error variance q:

$$q = Var(Z) - 2\sum_{i=1}^{N}\lambda_i \, Cov(Z(x_0)Z(x_i)) + \sum_{i=1}^{N}\sum_{j=1}^{N}\lambda_i\lambda_j \, Cov(Z(x_i)Z(x_j))$$

Notice that the square of the mean of Z cancels when summing these three components. More importantly, notice that $Var(Z(x_0))$, the variance of Z at the specific location, x_0, is now written as $Var(Z)$ which is location independent. This is a fundamental aspect of spatial analysis known as the **intrinsic hypothesis,** which holds that the variance of Z is everywhere constant in space, such that $Var(Z(x_i))$ is equal to $Var(Z(x_j))$ is equal to $Var(Z)$. This notion is also referred to as **second order stationarity** because variance is the second order moment of Z.

If the error variance, q, is differentiated with respect to the unknowns, λ, then N equations are obtained enabling a solution for these unknowns:

$$\frac{\delta q}{\delta \lambda} = \sum_{j=1}^{N}\lambda_j \, Cov(Z(x_i)Z(x_j)) = Cov(Z(x_0)Z(x_i)), i = 1, 2, \ldots, N$$

In fact, this solution is that for **simple kriging.** In this method, the previous equation given for the kriging estimator is modified slightly:

$$Z^*(x_0) = \bar{Z} + \sum_{i=1}^{N}\lambda_i(Z(x_i) - \bar{Z}); \quad \bar{Z} = mean \, of \, Z$$

Because there is no constraint on the sum of the weights, λ, this modified equation for kriging is necessary to assure **unbiased estimation.** That is, the mean of all estimated values, Z^*, is equal to the mean of the original data, Z.

Another form of kriging, **ordinary kriging,** does include a constraint such that the N weights sum to one. Because this constraint is included, and because unbiased estimation is the result, then the kriging estimator is that which was given before, $Z^*(x_0) = \sum \lambda Z(x_i)$. Constrained optimization using Lagrangian mathematics is used to rewrite error variance, q, to include the constraint:

$$q(\lambda,\mu) = Var(Z) - 2\sum_{i=1}^{N}\lambda_i \, Cov(Z(x_0)Z(x_i))$$

$$+ \sum_{i=1}^{N}\sum_{j=1}^{N}\lambda_i\lambda_j \, Cov(Z(x_i)Z(x_j)) - 2\mu\left(\sum_{i=1}^{N}\lambda_i - 1\right)$$

The variable, μ, is known as the **Lagrangian multiplier** and provides the constraint on the weights. Differentiating error variance, q, first with respect to λ, then with respect to μ, yields $N + 1$ equations used to solve for the N weights plus the one additional value, the Lagrangian multiplier:

$$\frac{\delta q}{\delta \lambda} = \sum_{j=1}^{N} \lambda_j Cov(Z(x_i)Z(x_j)) = Cov(Z(x_0)Z(x_i)),$$

$$i = 1, 2, \ldots, N; \quad \frac{\delta q}{\delta \mu} = \sum_{i=1}^{N} \lambda_i = 1$$

Before concluding this section, the notion of **kriging variance** is introduced. This term is synonymous with **estimation variance,** and is also known as **error variance.** Kriging yields an estimate of its error at each spatial location. Recall the expression for squared error, q:

$$q = Var(Z) - 2\sum_{i=1}^{N} \lambda_i Cov(Z(x_0)Z(x_i)) + \sum_{i=1}^{N}\sum_{j=1}^{N} \lambda_i\lambda_j Cov(Z(x_i)Z(x_j))$$

Kriging variance is actually the solution of this expression:

$$q = Sill - \sum_{i=1}^{N} \lambda_i Cov(Z(x_0)Z(x_i)) + \mu$$

This further explains why the sill of the variogram is often taken to be the variance of Z, and further why the calculation of the variogram is one-half the average squared difference as a function of h so that the sill will approximately or exactly match sample variance.

Kriging variance is often criticized for not being a true estimate of error. Rather, that it is best used to judge how close the nearest N data locations are to each estimation location. The second term of the kriging variance equation is a function of spatial autocovariance between each of the N nearest data locations and the estimation location. The closer at least one of these data locations is to the estimation location, the larger in value will be this second term, and the smaller will be the kriging variance. But, if we accept that the closer an estimation location is to a data location, then the more likely its true value is to be similar to the value at the nearby location and kriging variance can then be accepted as a valid measure of estimation error.

5.4.1   Covariance is Obtained From the Variogram

In the development of the kriging equations, those that are used to solve for the weights (and Lagrangian multiplier in ordinary kriging), the notion of covariance is essential, therefore must be somehow defined. In fact, spatial covariance is inversely proportional to the variogram:

$$Cov(h) = Sill - \gamma(h)$$

for which Sill is the sill of the variogram. Covariance as written in the kriging equations,

$$Cov(Z(x_i)Z(x_j))$$

is actually $Cov(h)$, based on a distance, h, that separates the two spatial locations, x_i and x_j. Likewise, the covariance, $Cov(Z(x_0)Z(x_i))$, is a function of distance that separates the two locations, x_0 and x_i.

Consequently, we must be able to define spatial correlation within the kriging algorithm as a function of separation distance, h. We do so using several different **variogram models.** These are functions that represent the idealized shape of the variogram for a particular set of spatial data. One strict rule for a variogram model is that it be a negative, semi-definite function such that it yields positive, or zero values of covariance (variance and covariance, being second order statistical moments, must always be positive or zero). Two common models are:

1. Spherical

$$\gamma(h) = \begin{array}{l} C0 + C\left(\dfrac{1.5h}{R} - \dfrac{0.5h^3}{R^3}\right), h < R \\[2mm] Sill, h \geq R \end{array}$$

   for which $C0$ is the nugget value and C is equal to the difference, $Sill - C0$.

2. Exponential

$$\gamma(h) = C0 + C\left(1 - EXP\left(-\frac{|h|}{R}\right)\right)$$

in which $C0$, the nugget value, C, the difference between sill and nugget, and R, the range, represent consistent symbolism with the spherical model. There are several other valid variogram models such as the Gaussian and linear models. Again, the strict requirement for validity is negative semi-definiteness.

Recalling Figure 5.3, variograms A, B, and D reveal a spherical spatial behavior, linear near the origin, becoming nonlinear as the sill is approached, then becoming constant thereafter. The program, *Visual_Data*, attempts the fit of a spherical model to the computed variogram. Based on an arbitrary programming choice and not on scientific principle, this model is chosen to have a sill that is equal to sample variance, a nugget that is equal to one-half the value of the first variogram increment, and a range that is equal to lag distance at which the computed variogram first exceeds the sill. The sole purpose of this example model fit is to enable the program user to decide what values of sill, nugget, and range are appropriate for representing the spatial correlation of a set of spatial data. Figure 5.3C is a variogram that exhibits exponential growth for which the exponential model is more appropriate.

5.5   The Practice of Kriging

Theory expressed on paper is one thing. Implementing theory in practice is another. Often, implementation requires a certain artistic license when "bending" theory to make it work. Kriging is certainly no exception.

Critical concerns when practicing kriging are:

- normality of spatial data;
- second order stationarity of spatial data;
- the variogram model;

- the design of the grid;
- the number of nearest, neighboring values to use during estimation;
- the size of the estimation window within which to find these nearest, neighboring locations;
- sample support, punctual or block;
- the necessity of applying a transform to the data, such as log-normal or indicator;
- the need to model anisotropic spatial covariance.

These considerations are discussed separately.

5.5.1   On the Normality of the Spatial Data

Often in the practice of kriging, one presumes that the spatial data are normally distributed. In fact, it is argued herein based on theory presented in Carr (1995, p. 154) that normality is a requirement only for the N nearest neighboring locations used in estimation. As long as N is kept small, for instance five or ten, then normality is a good assumption (recall Chapter 2 wherein it was shown that, in the chi-square test, the smaller the data set, the more likely it is that the normal distribution model represents the data distribution). Some books (e.g., Kitanidis, 1997, p. 71) recommend using the entire data set for N. Not only does this yield a very large matrix that must be inverted, setting N this large also demands that the entire set of spatial data be represented well by a normal distribution model. Recall from Chapter 2 that the larger the data set, the less likely it is that the normal distribution model is representative. Kriging is quite robust (insensitive to distribution assumptions) provided N is kept small.

5.5.2   On Second Order Stationarity

Like the question of normality, we must consider if the requirement of second order stationarity applies to the entire set of spatial data, or only to the subset of N nearest neighbors used in estimation. In the case of second order stationarity, there are several texts, most notably that of Journel and Huijbregts (1978) that discuss **quasi-stationarity** only for the N nearest neighboring locations. Once again, as long as N is kept relatively small, the assumption of quasi-stationarity is sound.

5.5.3   On the Variogram Model

This is perhaps the most crucial consideration, given the fact that the variogram model is the sole representative of the spatial autocorrelation inherent to the spatial data to which kriging is applied. Nevertheless, the variogram model may be adjusted in several ways to change the visual appearance of the final map. For example:

- increasing the nugget causes smoothing during kriging, resulting in a map having a smoother appearance (data variability is de-emphasized);
- decreasing the nugget has the opposite affect, minimizing smoothing thus accentuating data variability;
- increasing the range has the same effect as increasing the nugget; decreasing the range has the same effect as decreasing the nugget;

- increasing the sill has the same effect as decreasing the nugget; likewise, decreasing the sill has the same effect as increasing the nugget.

Of course, the application of kriging may have more rigorous, quantitative objectives. In this case, weighted least squares regression may be used to obtain parameters for a variogram model (e.g., Cressie, 1991).

5.5.4   On the Design of the Grid

Kriging may be used to estimate spatial data at nodes of a regular grid. This is a necessary prelude to automated contour line drawing. Grid design is accomplished by specifying values for several parameters. Resolution is a key consideration. The total number of rows and columns (and elevations in a three-dimensional model), combined with the physical spacing between rows and columns (and elevations) defines resolution. The fewer the rows, columns, and elevations, and consequently the greater the distance between these, the coarser is the grid resolution. The greater the rows, columns, and elevations, and consequently the lesser the distance between these, the finer is the grid resolution.

Resolution governs how well the original spatial data are reproduced in the final map. The finer the grid, the better is the reproduction. In other words, the coarser the grid is, the more the data are smoothed. The result of smoothing is the loss of extreme data value, high and low, with a greater emphasis on the mean data value. The finer the grid, the better is the reproduction of higher and lower data values.

5.5.5   More on the Number of Nearest Neighbors Used for Estimation

Discussion is presented in sections 5.5.1 and 5.5.2 arguing against the use of a large number of nearest neighbors when performing kriging. Additionally, this number, N, controls the size of the matrix system that must be inverted. The larger N is, the more memory that is required to form this matrix system, moreover the more substantial is the amount of time necessary to invert this system in the process of obtaining a solution for the weights. McCarn and Carr (1992) further show that the larger this matrix system is, the more severe is the roundoff error during equation solution. For this reason, the program, *Visual_Data* employs iterative improvement during Gaussian elimination when solving for the kriging weights.

Using a smaller number of nearest neighboring locations during kriging can create artificial discontinuities in the final map when data values change rapidly over space. Using a larger number of nearest neighbors will tend to smooth these data transitions, resulting in a fewer number of false discontinuities. This highlights the primary advantage of larger N. Visual examples presented subsequently underscore the differences between smaller and larger N.

5.5.6   On the Size of the Search Window (and the Type of Search Strategy)

Allied closely with the number of nearest neighbors is the size of the search window to use when locating these neighboring locations. If this window size is set relatively small, then N is the maximum number of spatial locations used during estimation. In many computer programs, including *Visual_Data*, the minimum number of locations required to proceed with kriging is two, and may range up to N, with any number between two and N possible. If the window size is set very large, tantamount to infinitely large, then N is always the number of nearest neighbors used during estimation. When using kriging for gridding as a prelude to contouring, the window size must be set very large to ensure that every point in the grid is associated with an estimated value. Otherwise, if fewer than two spatial locations are found within the search window centered about a grid location, an estimate will not be computed for the location.

Search windows are circles centered about a point to be estimated. A data analyst must specify the physical size of the radius of this window. This radius governs the size of the search window, smaller or larger. Additionally, the search strategy within this circular window must be considered. A general search strategy locates the N closest data locations, regardless of where in the circular window they occur. A quadrant strategy divides the circle into four equal segments, then locates an equal number of neighboring locations in each segment. The intent is to balance the neighboring locations equally surrounding the location at which an estimate is desired. A more precise refinement involves an octant search strategy based on subdividing the circular window into eight equal segments and finding an equal number of neighboring locations in each of these segments.

5.5.7   On the Concept of Sample Support: Punctual or Block

Spatial data represent values at precise spatial locations, or they represent an average value over an area or volume. Likewise, estimated values from kriging are taken to represent values at precise locations, or averages over areas or volumes. If a value is representative only of a precise spatial location, a **punctual,** or **point support** is implied. Otherwise, if a value represents an average over an area or volume, a **block support** is implied.

Kriging may be applied to accommodate punctual support (punctual kriging) or block support (block kriging). The matrix system presented earlier for kriging is that which yields a punctual estimate. In block kriging, because an estimate represents an average over an area or volume, the covariance terms in the right-hand side vector of the kriging matrix system, that which is based on the distances between nearest neighboring spatial locations and the location at which an estimate is desired, are integrated over the area or volume. The matrix system therefore becomes

$$\frac{\delta q}{\delta \lambda} = \sum_{j=1}^{N} \lambda_j Cov(Z(x_i)Z(x_j)) = \int_{A;V} Cov(Z(x_0)Z(x_i)),$$

$$i = 1, 2, \ldots, N; \frac{\delta q}{\delta \mu} = \sum_{i=1}^{N} \lambda_i = 1$$

Of course, digital computers cannot integrate. Consequently, the program, *Visual_Data*, employs a Gauss Quadrature numerical integration solution for block kriging (Carr, 1995, provides details).

5.5.8   On the Need for a Data Transform

Many practitioners assume that if the sample data are not represented well by a normal distribution model, then they cannot apply kriging. Or, they assume that a data transform is necessarily applied to the data before kriging.

This is true only when the number of nearest neighbors, N, used in kriging is set equal to the entire data set. Otherwise, restricting the search window and keeping the number of nearest neighboring locations small renders kriging more robust, or insensitive to data distribution.

Data transforms may be considered for reasons other than conformity to distribution model. Should the data be log-normally distributed, then a natural log transformation is usefully applied to the data both to estimate the variogram and for kriging. If such a transform is used, then a nonlinear form of kriging is obtained:

$$Z^*_{\ln}(x_0) = \overline{Z_{\ln}} + \sum_{i=1}^{N} \lambda_i\, f(Z(x_i)); \quad f(Z(x_i)) = \ln(Z(x_i)) - \overline{Z_{\ln}}$$

This is known as **log-normal kriging.** This form of kriging should be applied only to data that are represented well by the log-normal distribution model. For such data, a nonlinear estimator is more appropriate. The difficulty with log-normal kriging is in the back transform:

$$Z^*(x_0) = e^{Y^*_{\ln}(x_0) + \frac{\sigma}{2}}; \quad \sigma = Sill_{\ln} - \sum_{i=1}^{N} \lambda_i\, Cov_{\ln}(Z(x_0)Z(X_i)) - \mu$$

that is biased unless the data, Z, conform precisely to a log-normal distribution model.

Another useful transform involves a nonparametric (robust) dummy coding known as **indicator coding.** Given data, Z, having a range, $R = b - a$, then if an arbitrary value, c, is chosen such that

$$a < c < b$$

then an indicator transform is obtained as

$$i(x) = 0 \text{ if } Z(x) > c$$

$$i(x) = 1 \text{ if } Z(x) \le c$$

This simple transform results in a binary (two-valued; 0 or 1) function that is nonparametric because the notion of a distribution model is not relevant when data take on only two possible values.

Indicator transforms are useful for:

- minimizing the influence of unusual data values in variogram calculation and kriging;
- muting the influence of noise when calculating the variogram;
- more precisely resolving the range of spatial autocorrelation (in this case, setting c equal to the median value of Z often results in the best looking variograms);
- highlighting unique intervals of data values, for instance higher or lower data values; this application is analogous to binary contrast stretching in digital image processing;
- mapping the probability that Z is less than or equal to a particular value, c. Notice that i, as defined, can be viewed as a probability.

5.5.9   On Anisotropic Spatial Autocorrelation Modeling

Spatial autocorrelation may vary directionally. This type of spatial behavior can be resolved using directional variograms. Computing variograms typically in four directions, east-west, northeast-southwest, north-south, and northwest-southeast, moreover allowing 45 degree windows for each direction, yields variograms that may, or may not, have equal ranges. If the ranges of these directional variograms are approximately equal, then isotropic spatial autocorrelation is inferred. If, on the other hand, the ranges differ, then anisotropic spatial autocorrelation is assumed.

The anisotropic behavior is accommodated in kriging in the search strategy, by weighting neighboring locations more in one direction than in another. The direction favored is that associated with the largest variogram range. Additionally, kriging requires the ratio of the largest variogram range divided by the smallest variogram range. Knowing the direction associated with the largest variogram range (major axis) and this ratio defines an elliptical search window in kriging, rather than a circular search window that was described earlier.

5.6   Visualizations

So many considerations and decisions are necessary when practicing kriging. Consequently, two different practitioners applying kriging to the same spatial data will likely create two different maps. An interesting experiment was that of Englund (1990), wherein the same data set was given to twelve different analysts. The outcome was twelve different pictures! Each researcher chose a different approach in the practice of kriging. Some chose to use log-normal kriging, others chose ordinary kriging, each chose a unique set of variogram parameters, and so on. These decisions resulted in different pictures of the actual spatial data.

In this section, the same data set is treated using a different set of parameters in kriging to underscore the visual impact of each parameter. Some parameters have a subtle visual impact, whereas other parameters have a more substantial visual impact.

A simulated spatial phenomenon is used for demonstration. Simulation is presented in this text in detail in Chapter 7. Pictures of this simulation and its associated variogram are shown to begin this experiment:

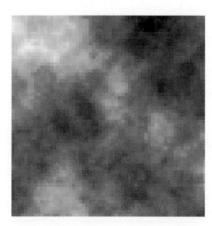
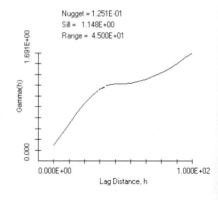

The first experiment involves the application of kriging in varying attempts to reproduce this picture by varying the resolution of the grid. In each experiment, the variogram model used is a spherical one for which the nugget value is set to zero, the sill value is set to 1.2, and the range is set to 30.0. The number of nearest neighboring locations, N, used in kriging is 5. Grid resolution is now varied as follows: A. 100 rows $\times$ 100 columns, a resolution that matches the original simulation; B. 60 rows $\times$ 60 columns, a coarser resolution; and C. 20 rows $\times$ 20 columns, the coarsest resolution:

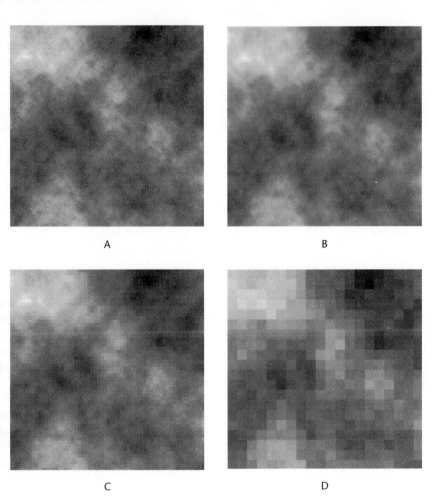

A   B

C   D

▲ FIGURE 5.4
Three kriging experiments demonstrating the effect of grid resolution. Starting with the upper, left image: A. original simulation; B. kriging using a 100 $\times$ 100 grid, the same resolution as for the original simulation; C. kriging using a 60 $\times$ 60 grid; and D. kriging using a 20 $\times$ 20 grid.

Grid resolution affects our ability to visualize spatial data, but it is important to realize that we cannot see what is not there. In other words, it is the spatial data themselves that actually control what we can see. Grid resolution simply impacts how well we can preserve nuances in these data. It is pointless, though, to design grid resolution finer than the resolution of the spatial data because we cannot visualize, through kriging, what is not in the spatial data. Suppose we use a grid resolution of 200 x 200 applied to the simulated data:

 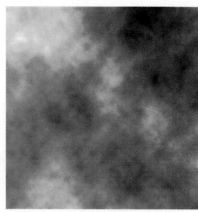 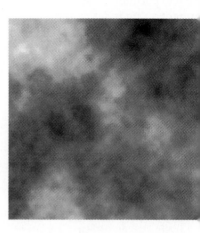

▲ FIGURE 5.5

An experiment to show what happens when grid resolution exceeds the data resolution. The original simulation (left most image) is estimated using a 200 × 200 grid (middle image). The result from kriging using a 100 × 100 grid (right most image) is shown for comparison. Both kriging results are comparable. In other words, although a finer grid is used, the 200 × 200 grid does not represent twice the resolution as the data (left hand image), but instead has the same resolution as the actual data.

An experiment is presented to demonstrate the smoothing effect of the variogram nugget. Three estimation experiments are presented using a 60 × 60 grid and varying the size of the nugget from one-third the sill value up to the sill value. The larger the value of the nugget, the more equal are the weights used in kriging. When the nugget is equal to the sill value, kriging becomes a local average estimator because each weight is equal to $1/N$, and N is the number of nearest neighboring locations used in the estimation process (shown at the top of the next page).

Another experiment shows the influence of N, the number of nearest neighboring data locations used in kriging. Again using a 60 × 60 grid, setting $N = 5$ is contrasted with $N = 75$ (shown at the bottom of the next page). This experiment visually demonstrates the influence of N in kriging. In this case, the visual results are almost identical. Yet, increasing N from 5 to 75 resulted in a twenty-fold increase in the amount of computer time required to complete the grid. Given that the visual results are almost identical, the substantial increase in computer time does not seem warranted. Moreover, this experiment underscores statements forwarded earlier arguing for the use of smaller N.

Block kriging is contrasted with punctual kriging to show its visual impact on estimation. All foregoing experiments were developed using punctual, ordinary kriging. Block, ordinary kriging is expected to more severely smooth the data in comparison to punctual kriging because of the numerical integration employed in the calculation of the right-hand side vector of covariances based on separation distances between the N nearest neighboring data locations and the location at which an estimate is desired.

Finally, several experiments are presented to demonstrate the indicator transform. This simulation has a mean of zero and variance of one. The simulation algorithm is designed to yield spatially correlated numbers that are

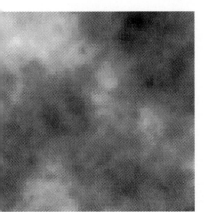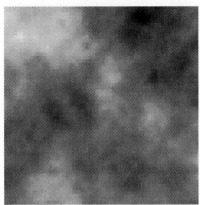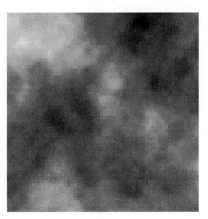

▲ FIGURE 5.6

Three experiments to demonstrate the visual impact of the variogram nugget. All four images were developed using kriging and a 60 × 60 grid. The top image used a nugget value of zero and is shown for comparison. The bottom three images, left to right, were obtained using a nugget value equal to one-third the sill value, two-thirds the sill value, and the sill value respectively. All are comparable in visual appearance. All three are smoother (more blurry) in comparison to the top most image, and blurriness increases with nugget, from left to right.

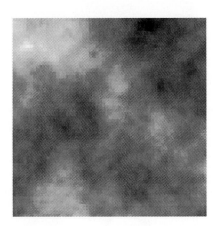

◀ FIGURE 5.7

An experiment to demonstrate the visual influence of the number of nearest neighboring data locations used in kriging. A 60 × 60 grid and nugget value equal to zero was used to obtain each image. The left-hand image was obtained using the $N = 5$ nearest data locations, whereas the right-hand image was obtained using the $N = 75$ nearest data locations.

represented well by a normal distribution model. The median should consequently be zero. When using the indicator transform, the best variograms are often obtained when the arbitrary cut off, c, is close in value to the median data value (Figure 5.10). Values in the simulation range between -3 and $3.$ The experiments attempted now in application to the simulated data involve $c = -1.5$, $c = 0.0$ (median), and $c = 1.5$.

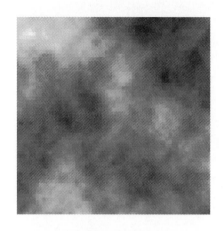

▶ **FIGURE 5.8**
A comparison of punctual, ordinary kriging (left image) to block, ordinary kriging (right image). A 60×60 grid was used to obtain each image. Variogram parameters were identical. Notice that the block kriging result is more blurry indicating a higher degree of smoothing.

Recall the indicator transform: $i(x) = 1$ if $Z(x) \leq c$; or $i(x) = 0$ if $Z(x) > c.$ One way to think of the indicator transform is that i is larger in value when Z is smaller in value. By this characterization, we see that i is an inverse function of Z. This realization aids the interpretation of the following visualizations.

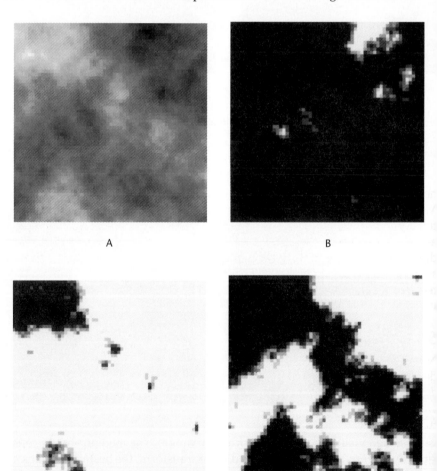

A   B

▶ **FIGURE 5.9**
Three indicator transform experiments. A. ordinary kriging using a 60×60 grid for comparison; B. indicator transform using $c = -1.5$; C. indicator transform using $c = 1.5$; and D. indicator kriging using $c = $ median $= 0.0$. Notice that the indicator transform yields an inverse picture of the actual data; black pixels in the indicator transform images represent regions greater in value than c.

C   D

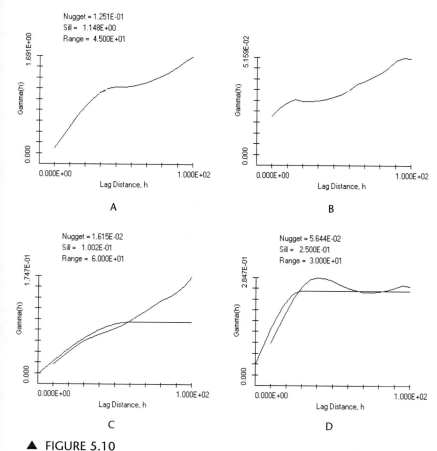

▲ FIGURE 5.10

Four variograms. A. variogram of the original simulation; B. indicator variogram for $c = -1.5$; C. indicator variogram for $c = 1.5$; and D. indicator variogram for $c = 0.0$ (median). Notice how the variogram changes morphologically and parametrically with cutoff, c.

Not only does the indicator transform yield a picture of a spatial phenomenon with an appearance that changes with c, but the variograms differ as well. Morphologically, the best variogram is obtained when c is set equal to the data median. In this case, exactly one-half the indicator values, i, are zero, and the other half equal one. This parity in frequency provides a more equal distribution of ones and zeros over space, resulting in the more ideally shaped variogram. Such an indicator variogram is useful, for instance, for more precisely determining the correct range of the spatial structure to use in variogram modeling.

5.7.   Application to the *Nevada_Landsat_Data*

Now that the optical influences of variogram and kriging parameters are demonstrated, this knowledge is applied to the modeling of the *Nevada_Landsat_data*. Up to now, the *Nevada_Landsat_6x* data have been analyzed. This data set was developed by resampling the original 600 pixel × 600 row extraction of a full resolution Landsat TM image by taking every sixth pixel and every sixth row. This resampling compressed the original 360,000 pixels to yield 10,000 pixels, a more manageable data set for many of the analyses presented in this text.

Other resamplings are presented on the CD-ROM of these same data. These represent 2x and 4x resamplings, taking every other pixel and row and every fourth pixel and row. Additionally, to mimic more closely how data are collected by a human observer in the field, clustered data were obtained of this image by selecting smaller groups of pixels representing: dense vegetation, sparse vegetation, water, different rock types, alluvium, and shorelines. All of these data sets are analyzed using kriging to demonstrate the different resultant visualizations of the same spatial phenomenon.

Recall from Chapter 2 that none of the variables associated with the *Nevada_Landsat* data set has a distribution that is represented well by the normal distribution model. We will proceed with kriging in spite of this by keeping the number, N, of nearest neighboring locations small when estimating. Because the focus in this chapter is on univariate spatial relationships, and further because variograms are presented earlier only for visible blue reflectance, it is this variable that will be treated in kriging. The intent of this presentation is the demonstration of kriging applied to an actual set of data and the visualization of results.

In the first set of experiments, kriging is applied to data that are resamplings of the original Landsat TM image. Three data sets are used: *Nevada_Landsat_2x*, *Nevada_Landsat_4x*, and *Nevada_Landsat_6x*. The portions of the names, 2x, 4x, and 6x, represent the resampling interval. These three data sets represent differing resolutions of the same spatial phenomenon. Each full resolution, Landsat TM pixel represents a region on Earth that is 30 m by 30 m in size. The 2x data set consequently yields pixels having 60 m resolution; pixels in the 4x data set have 120 m resolution; and those in the 6x data set have 180 m resolution. Visualizations are used to see how much of the original data is captured by these differing resolutions.

Kriging can only reproduce spatial information that is captured in the data to which it is applied. Indeed, these data should control, not be controlled by the analysis. Consequently, when examining the visualizations presented in Figure 5.11, notice that only those spatial features as large or larger than the resampled resolution are reproduced. The coarser the resampling is, the more severe is the loss of finer detail in the image. This finer detail is known as **high frequency** information, or information that changes over a very small distance in space. The larger features, the coarser detail, represent **low frequency** information, or information that is prominent over a larger spatial area. Kriging is a smoothing operation, which means that it will remove, or filter, some of the higher frequency information from spatial data. It is important to further consider that the manner in which we sample a spatial phenomenon can result in the loss of a particular amount of higher frequency information.

In particular, when geologic data are collected in the field, a regular sampling pattern may, or may not be used. Often, field conditions are not conducive to sampling on a regular grid. Barriers may prohibit such regular sampling. Instead, data may be obtained in groups, or clusters. Kriging can still be applied to such data, yet the picture (map) obtained of the spatial phenomenon may be distorted owing to this clustered sampling.

For example, in Chapter 4, the data set, *Nevada_Landsat* clustered was used to test the multivariate relationships among several variables. Applying kriging to these data in the estimation of visible blue reflectance yields the image shown on the bottom of the next page (Figure 5.12). Notice that the kriging model is good inside the cluster, but does not yield a valid representation of the actual spatial phenomenon outside the cluster. Once again, kriging can only reproduce information captured in the data to which it is applied.

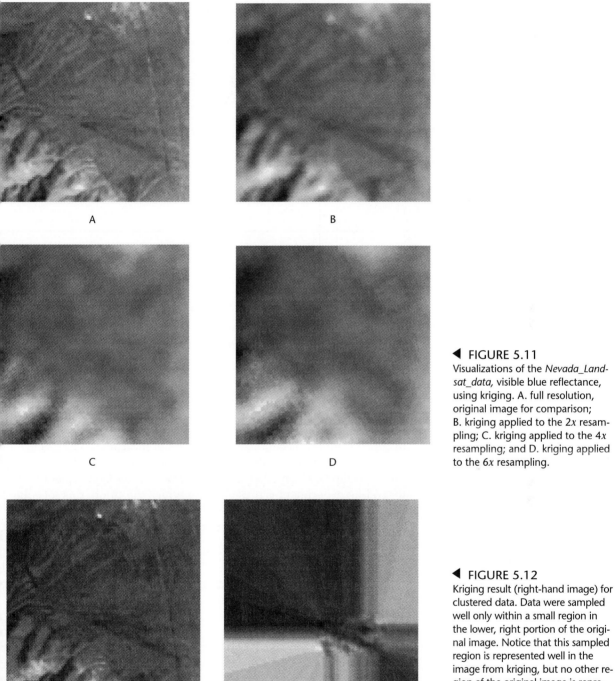

A

B

C

D

◀ **FIGURE 5.11**
Visualizations of the *Nevada_Land-sat_data,* visible blue reflectance, using kriging. A. full resolution, original image for comparison; B. kriging applied to the 2*x* resampling; C. kriging applied to the 4*x* resampling; and D. kriging applied to the 6*x* resampling.

◀ **FIGURE 5.12**
Kriging result (right-hand image) for clustered data. Data were sampled well only within a small region in the lower, right portion of the original image. Notice that this sampled region is represented well in the image from kriging, but no other region of the original image is represented well due to a lack of data.

The application of kriging to the 2*x*, 4*x*, and 6*x*, resampled data relied on the same variogram, modeled as a nested combination of two spherical models. The nested model is that which is shown previously. The nested model is an attempt to fit the variogram that is obtained for the 6*x* resampled data (Figure 5.2). That this model is validly applied for the other resamplings is suggested by the variogram obtained for the 4x re sampled data:

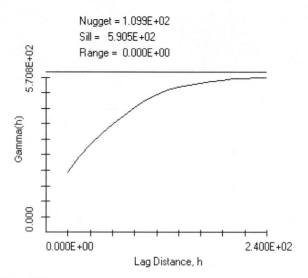

▲ FIGURE 5.13
Variogram for the *Nevada_Landsat_4x* data set, a 4x resampling of the original Landsat image.

Notice that this variogram is almost identical to that for the 6x resam pled data. Nugget, sill, range, and nested appearance are comparable. Wha does this suggest about the self-affine fractal nature of these Landsat data?

In concluding this section, an application of indicator coding is used to demonstrate the isolation in these data of higher (and lower) regions of visi ble blue reflectance. The median blue reflectance is 155. Consequently, an in dicator coding experiment is presented for which the cutoff, c, is set equal to the median blue reflectance.

The variogram for this indicator transform more closely resembles an exponential autocovariance than a nested structure:

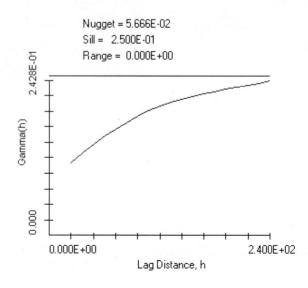

Modeling this autocovariance using a single exponential model, nugget = 0.06, sill = 0.25, and range = 240, resulted in the following map:

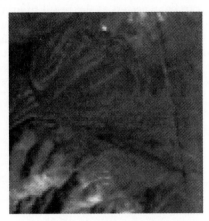 

▲ FIGURE 5.14
An indicator coding map (right-hand image) for cutoff = 155 (median blue reflectance). Recall that an indicator coding gives an inverse relationship; values in the original image (left-hand image) that exceed 155 are darker in the indicator coding map. The indicator kriging map reveals the spatial pattern of higher blue reflectances.

5.8   *M*-Kriging

Extensive information was presented in Chapter 3 on m-regression, a robust form of least squares. Kriging is also a least squares approach, and the m-regression algorithm can consequently be implemented within it to yield a more robust form of estimation. This is not a new presentation. Robust kriging is discussed at length in Chiles and Delfiner (1999), based on previous work by Hawkins and Cressie (1984).

The algorithm is identical to that presented earlier for regression, written now for kriging. This is a simplified method from what is described by Hawkins and Cressie (1984). Eventually, results from ordinary kriging are compared to those from *M*-kriging to judge the robustness of the former method when the kriging neighborhood, *N*, is small. Simplified *M*-Kriging:

1. compute a variogram for the data, Z, and decide on an appropriate model for it;
2. apply kriging in **cross-validation mode** to develop the following deviation, $Z(x) - Z^*(x)$, at each data location;
3. compute the median value of these [absolute] deviations; call this MAD;
4. estimate the standard deviation, σ, of these deviations as 1.483MAD;
5. adjust the original data, Z, as follows:

$$Z(x) = Z^*(x)+1.5\sigma, \text{ if } (Z(x) - Z^*(x)) > 1.5\sigma$$

$$Z(x) = \text{unchanged}, -1.5\sigma < (Z(x)-Z^*(x)) < 1.5\sigma$$

$$Z(x) = Z^*(x)-1.5\sigma, \text{ if } (Z(x) - Z^*(x)) < -1.5\sigma$$

Unlike M-Regression, steps 2 through 5 are completed only once.

One of the data sets used by Englund (1990) in an experiment to determine the variability in spatial data analysis that can be expected as the result of arbitrary decisions on the part of analysts is analyzed using ordinary, punctual and M-punctual kriging. This data set, Area1.dat, is found on the CD-ROM in the folder, \data\othrdata\. Applying ordinary, punctual kriging to these data in cross-validation mode using the N = 5 nearest neighboring data locations yields the following scatter diagram of true (*x*-axis) versus estimated (*y*-axis) value (left diagram):

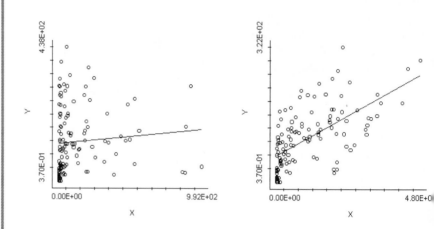

Using an M-Kriging approach to filter large deviations from the data yields the scatter diagram shown above on the right. In this case, estimation is able to reproduce higher and lower data value with less error.

How much of a visual impact does *M*-Kriging have in comparison to ordinary kriging? This is tested in application to the *Nevada_Landsat_6x* data set (left image = ordinary, punctual kriging; right image = M-Kriging):

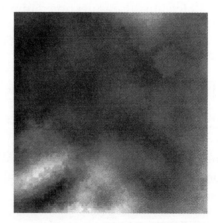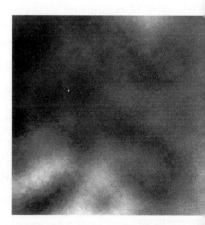

These two images are not appreciably different. This visually supports the notion that kriging is more robust if the number of closest neighboring data locations, *N*, is kept small. In this application, ordinary punctual kriging (left image) was implemented using the five closest neighboring locations. *M*-kriging, right image, a more robust form of kriging, yielded approximately the same result.

.8.1   Cross-Validation

'ross-validation, known also as a "leaving-one-out" procedure, or colloqui-
lly as "jack-knifing," is an implementation of kriging for estimating a value,
.*(x), based on the closest N nearest neighbors to x, without using the actual
ata value at the location, x. This enables the calculation of an actual error
not a theoretical error as is the case with kriging variance), $E = Z(x) -$
.*(x). The following statistics and visualizations from cross-validation pro-
ide useful insight to kriging performance:

- the mean value of E; this quantity should be close to zero for unbi-
  asedness;
- the histogram of errors, E, useful for judging the conformity of these
  errors to a normal distribution model; deviation from a normal distribution model may indicate some systematic bias in the estimation, perhaps due to a trend in the data values over a particular direction;
- a plot of $Z(x)$ versus $Z^*(x)$ shows how closely estimated values match
  true values; can be useful when comparing different kriging approaches, such as simple kriging versus ordinary kriging; or when comparing
  several different variogram models for performance;
- filters large deviations in Z during M-Kriging.

Cross-validation is additionally useful for computing a jack-knife statis-
ic for each estimated value. This statistic is computed as

$$jk = \frac{Z(x) - Z^*(x)}{\sqrt{\sigma_K^2}}$$

[the jack-knife statistic, jk, is equal to the difference between the estimated
value and the true value; this difference divided by kriging standard deviation, the square root of the kriging variance]

f the jack-knife statistic, jk, is less than -2, or greater than 2, then the differ-
nce between true and estimated value is more than 2 kriging standard devi-
tions, and we conclude that the true value, $Z(x)$, is a "rare event" with
espect to its nearest neighboring data locations. This does not indicate that
$Z(x)$ is in error and consequently should be removed. It does suggest that its
alue is spatially unusual and this may be significant.

5.9   How Do I Reproduce Results in This
Chapter, Or Analyze My Own Data . . .

5.9.1   . . . Using *Visual_Data?*

.9.1.1   Time Series Analysis

f the data are already in a file, then the first step involves opening this file in
he main program window. If the data are not in a file, they may be entered
n the upper box of the main program window using the keyboard. If the
lata are entered in this fashion, it is important to save these data to a file be-
ore proceeding to time series analysis. The Save (not the Save As) option is
used from the File menu for this purpose.

Once the data file is opened, click Tools, then move the cursor to "Time
Series Analysis: Slow Fourier Transform." A slow Fourier transform implements the classic integral formula for this type of time series analysis. A data
ile of any size may be analyzed. In contrast, if one uses a Fast Fourier Trans-
orm algorithm, the data file must be of a size, 2^x, a power of 2 in size. (If a

data stream is not a power of 2 in size, zeros are used to "pad" the stream u
to a power of 2 in size).

As a demonstration, data representing the thicknesses of varve sedi
ments are analyzed (Davis, 1986). These data are found on the CD-ROM i
the directory, \data\timedata\. This file is opened in the main window o
Visual_Data, then the Time Series tool is clicked. Several interactive question
then prompt the analyst for information: 1. do you want to determine the frac
tal dimension of the time series? In the case of the varve data, we wish to loo
for significant cycles present in these data that may be useful for understand
ing their cyclic nature; consequently, we respond NO for the analysis of fracta
dimension; 2. how many data values are on each line of your file? this questio:
shows that this tool can be adapted to a number of different data file format:
provided the file has the same number of values on each data line; in the cas
of the varve data, the answer is one (value per line); 3. of these, which one is t
be analyzed? coupled with the second question, this question shows the flexi
bility of this tool; the analyst selects which data value on each data line is to b
considered; in the case of the varve data, we again respond with one, becaus
there is only one value per line. Once the third question is responded to, th
following graphical results are displayed. The data are displayed:

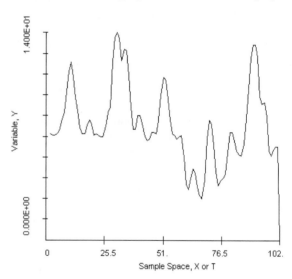

followed by a plot of the power spectrum (the squares of the Fourier transform)

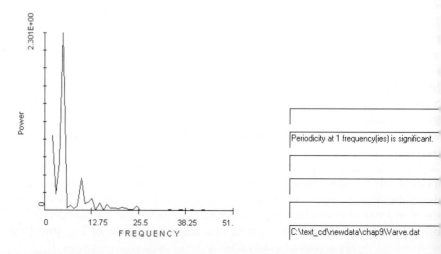

Notice that one frequency, five cycles (this information is found in the main window of *Visual_Data*, lower box), is identified. There are 101 varves, each representing an annual cycle, consequently a frequency of five corresponds to a wavelength of 20.2 years. Because at least one frequency is statistically significant, *Visual_Data* creates a filtered time series:

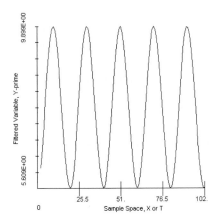

.9.1.2   Variogram Analysis

Data representing one, two, or three-dimensional space may be analyzed. For instance, suppose we wish to reproduce Figure 5.2. Once the data file, *Neva-a_Landsat_6x*, is opened (CD-ROM, folder \data\nvlandat\ we click Tools, then Variogram Analysis. Several interactive questions prompt the analyst for information: 1. has a data file been opened, yes or no? We respond YES; 2. enter the option for directional analysis: enter one for omnidirectional (an average over all spatial directions), two for one, user-specified direction, useful when we suspect that one direction will be dominant in our analysis, four for a four-directional analysis, E-W, NE-SW, N-S, and NW-SE, useful when we wish to investigate anisotropic spatial autocorrelation, yet we are unsure of the dominant direction, or eight for an eight-directional analysis to yield higher resolution in comparison to the four-directional analysis; select 1 omnidirectional, to reproduce Figure 5.2; 3. enter data transform option: 0 for no transform, 1 for log-normal transform, or two for indicator transform; if the indicator transform is selected, another question prompts the analyst to specify the value of the cut-off, c; we enter 0 to reproduce Figure 5.2; 4. define dimensionality of the data: enter one for one-dimensional; two for two-dimensional; or three for three-dimensional; enter 2 for these data; 5. enter the class size; the value used to obtain Figure 5.2 is 12; 6. Is your data in Simplified GeoEAS format? GeoEAS is a geostatistical (spatial) data analysis program developed by the United States Environmental Protection Agency (Englund and Sparks, 1988). This format is supported by the geostatistical software package, GSLIB (Deutsch and Journel, 1992; 1998). These are popular software packages and *Visual_Data* is designed to be compatible with their data files. In the case of the *Nevada_Landsat_6x* data, however, the GeoEAS format is not used, therefore we respond NO to this question; 7. How many data values are on each line of our data file? This feature provides flexibility with this tool in analyzing a wide variety of data file formats, provided the same number of data values are on each line; in the case of the *Nevada_Landsat_6x* data set, there are nine values per line: seven variables plus one *y* and one *x* coordinate, for a total of nine values; 8. what is the column position of the datum (the value, Z, to be analyzed)? In the case of visible blue reflectance in the *Nevada_Landsat_6x* file, we respond one to this question;

9. the position of the y-coordinate: we respond eight in the case of this file and 10. the position of the x-coordinate: we respond nine in the case of this file. Once this final question is responded to, *Visual_Data* completes the calculation of the variogram. This particular data file has 10,000 spatial locations requiring some computational time for the program to finish.

5.9.1.3 Kriging

To reproduce the kriging result for the *Nevada_Landsat_6x* data set (simply as one example), we start by opening this file in the main program window then we click Tools and move the cursor to Kriging. Four options then appear: 1. Create or Edit Grid Parameter file; 2. Create or Edit Variogram Parameter file; 3. Proceed to Estimation; or 4. Create a Bitmap image of kriging results. Before we can apply kriging to these data, we must create two files, one for grid parameters and the other for variogram parameters. *Visual_Data* is designed to guide users when creating these files.

For instance, we wish to model the variogram shown in Figure 5.2 using a single variogram model as follows:

Nugget = 100.0; Sill = 570.0; Range = 200.0;
Anisotropy Direction = 0.0;

Anisotropy Ratio = 1.0; we will use a general search option, large search window radius because we will grid these data; and we shall keep the number of nearest neighboring data locations used during estimation small, $N = 5$.

We click the option to Create or Edit Variogram Parameter File and the following window appears:

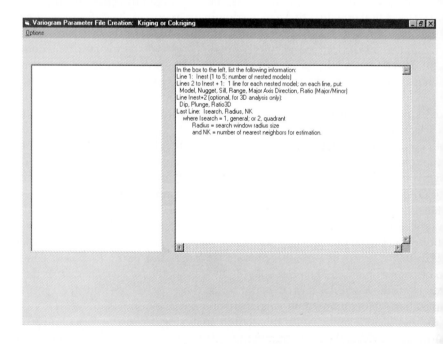

The box on the right hand side provides guidance for creating the file in the left hand box. Line 1 specifies the number of nested variogram models to be used. Nested models are a combination of two, or more, valid variogram models, the values from which are summed for a particular lag distance, h, to obtain the modeled value. In the case of visible blue reflectance for the *Nevada_Landsat_6x* data set, only one variogram model is used, so *Inest* is set equal to 1. The remaining parameters are defined as shown in the preceding paragram. The file looks like this when finished:

```
1
1 100. 570. 200. 0. 1.
1 100000. 5
```

Much discussion is presented earlier, though, on the nested aspect of the spatial autocorrelation exhibited by these data. A file for the two nested variogram models presented earlier in this chapter looks as follows:

```
2
1 111. 511. 120. 0. 1.
1   0. 82. 240. 0. 1.
1 100000. 5
```

This is, in fact, the variogram file that was used to obtain each of the images of blue reflectance for the *Nevada_Landsat 2x, 4x,* and *6x* data sets (the same variogram parameters were used for the three experiments).

As with variogram parameters, a similar help window is provided to aid analysts when creating the grid parameter file. This window is shown below:

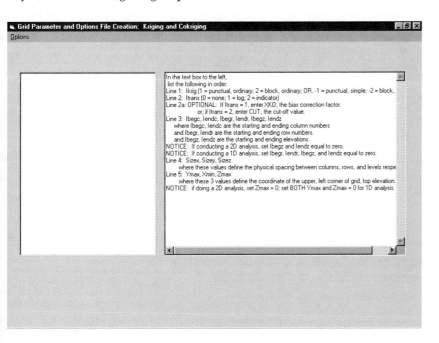

When contemplating the gridding of a set of data, this grid must necessarily be designed. A grid is a regular, rectangular or square shape, defined by a number of rows and a number of columns. In the kriging experiments applied to the *Nevada_Landsat_data,* for instance, a 100 row × 100 column grid was chosen. Once the number of columns and rows are decided upon (and elvations for three-dimensional gridding), the resolution of the grid is designed by specifying the physical distance between rows (sizey) and columns (sizex) (and possibly elevations, sizez). Finally, the coordinates of the upper left-hand corner of the grid are defined (Ymax, Xmin, Zmax). Note that grid ding proceeds TOP DOWN in the y-direction, and from LEFT TO RIGHT in the x-direction, and TOP DOWN if performing three-dimensional gridding in the elevation direction. These grid parameters are shown in the figure below.

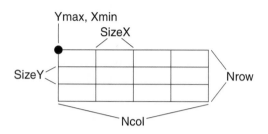

For the *Nevada_Landsat 2x, 4x* and *6x* data sets, the following grid was designed:

100 rows;   100 columns;   SizeX = 1.0 (1 pixel); SizeY = −1.0
Ymax = 0.5;   Xmin = 0.5;

Why was SizeY set to a negative number? Digital images start with row AT THE TOP, and increment downward. The data files for the *Nevada_Landsat* data have y-coordinates then that are smaller for the top of the image than for the bottom. Normally in kriging, gridding starts with row one at top and incre ments downward. This is consistent with the image. But, the kriging program is designed in the belief that, although row one is at the top, the MAXIMUM y coordinate is also at the top. *Visual_Data* computes the y-coordinate of each grid row as: $Ymax + 0.5(SizeY) - i$ Size Y, in which i is the row number. Notice the term, $0.5(SizeY)$. This appears such that kriging is used to obtain an estimate in the center of a grid cell, rather than at row/column intersections. By setting $Ymax = 0.5$ and $SizeY = -1$, then the two terms, $Ymax + 0.5(SizeY)$ cancel and we are left with $-i$ Size Y. Because SizeY is a negative number, $-i$ SizeY is positive. In this manner, the y-coordinate of each row grows larger as kriging proceeds downward. This is what we want to be consistent with our data file.

The file used for grid parameters is consequently:

```
1
0
1        100    1    100    0    0
1.       −1.    0.
0.5      0.5    0.
```

Notice that the beginning column number and beginning row number are each set equal to one. Suppose an analyst wishes to redo only a portion of their grid, say between rows 10 and 30, columns 40 and 60; then, the begin ning column number is set equal to 40, the ending column number is set equal to 60; similarly, 10 and 30, respectively, for the beginning and ending row numbers. Also notice that, whereas two-dimensional gridding is de signed, values of zero are required for the beginning and ending elevations. This is the signal detected by *Visual_Data* that two-dimensional gridding is chosen. Finally, notice that a value, 0, is entered for *Zmax*, even though two dimensional gridding is designed.

Once these two files are created, (from each of the file creation windows, when finished with file creation, click Options, then Save, to save the file; then, from the Options menu, a user must select the Return to Main Program item to go to any other option), kriging is started by clicking the Proceed to Estimation item. Several questions then prompt a user for further information: 1. has a data file been opened? 2. specify the grid parameter file; 3. specify the variogram parameter file; 4. is the data file in the Simplified GeoEAS format? 5. how many data values are on each line of the data file? 6. What is the column position of the datum, Z, to be kriged; 7. the column position of the y-coordinate; and 8. the column position of the x-coordinate. (Revisit Section 5.9.1.2 for proper responses to these queries).

At this point, *Visual_Data* scans the data file twice. The first scan simply counts the number of lines in the file (spatial data locations). This count is then used to define the size of arrays. The second scan reads the data into the arrays. Once the data are assigned to arrays, *Visual_Data* examines the data for duplicate samples, those having identical coordinates. Duplicate samples will cause a fatal error in kriging because the matrix system used to compute weights becomes indeterminate when duplicate samples make up the N nearest neighboring data locations. A user is alerted if duplicate samples are found.

After data access and duplicate sample scanning, a user is prompted to select cross-validation or gridding. Cross-validation is a method of kriging wherein an estimate is made at a location associated with a sample using the N nearest surrounding data locations, but not the one at which an estimate is computed. This gives an estimated value that can be compared to the actual value, useful when assessing overall kriging accuracy. Cross-validation is an important consideration in the next chapter on multivariate spatial analysis and a more detailed discussion of this method is deferred until then.

If the gridding option is selected in kriging, the program then prompts a user to select a filename for storing the numerical results from estimation. When gridding is finished, a user is given the option to create a bitmap image of the kriging results. It is these bitmap images that are shown throughout this chapter. Alternatively, users may create contour maps or three-dimensional grid perspectives of their kriging results. More detail on these visual display techniques is provided later in this text.

5.9.2   ... MATLAB

5.9.2.1   Time Series Analysis

Middleton (2000) provides a MATLAB function, pdg, for computing the power spectrum (periodogram) for a one-dimensional time series. An example repeats the analysis of the El Nino intensity data. The following steps are followed. After starting MATLAB, at the cursor prompt:

```
>>cd [CD-ROM drive letter]:\Time_Series\
>>load elnino.dat
>>X = elnino
```

Assuming the diskette for Middleton (2000) is in drive A:

```
>>cd A:
>>v = pdg(X, 256, 1., 'year')
```

This last step produces the graph on the next page. This power spectrum matches the analysis obtained using *Visual_Data* that was presented at the beginning of this chapter.

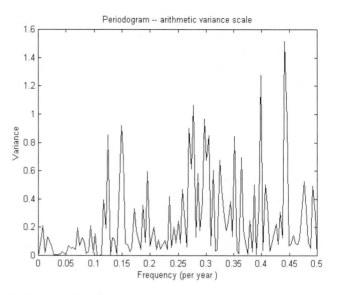

5.9.2.2   Variogram Analysis

Middleton (2000) presents a MATLAB algorithm for computing an omnidirection al variogram for regularly sampled data. This algorithm, semivxy, is modified and generalized for any type of geometric sampling and renamed semivxy_ (CD-ROM, MATLAB folder). Applying this algorithm in MATLAB requires the fol lowing steps. Suppose the *Nevada_Landsat_6x* data are analyzed, then:

1. Start MATLAB
2. At the >> prompt, enter:

```
>>cd [CD-ROM drive letter]:\Nevada_Landsat_data\
>>load Nevada_Landsat_6x.dat
>>W = Nevada_Landsat_6x
>>Z = W(:,1)
>>Y = W(:,8)
>>X = W(:,9)
>>cd [CD-ROM Drive Letter]:\MATLAB\
>>gam = semivxy_2(Z,Y,X,12., 100)
```

This final step eventually creates the following graph:

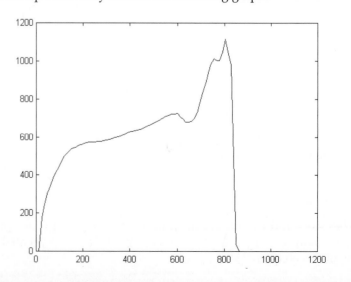

his represents better resolution for larger lag distances in comparison to re-
ilts presented earlier from *Visual_Data*. The sudden drop off in the end of
ie graph occurred because the limit of the data was reached (no pairs could
e found for larger lag increments). The portion of the variogram near the
rigin shows the distinctly nested aspect of this spatial autocorrelation.

.9.2.3   Kriging

Iiddleton (2000, p. 155) presents an algorithm, pkrige, for punctual, ordi-
ary kriging using MATLAB. This function, however, uses the entire data set
or the *N* nearest neighboring data locations, moreover yields an estimated
alue at only a single location. Consequently, this function was modified to
erform a regular gridding operation. Its application to the *Nevada_Land-
t_6x* data proceeds as follows:

1. start MATLAB; once in, the following symbol appears: >>; the following commands are then executed:
2. >>cd [CD-ROM drive letter]:*Nevada_Landsat_Data*\\
3. >>load *Nevada_Landsat_6x.dat*
4. >>W = *Nevada_Landsat_6x*
5. >>M(:,1) = W(:,9)   'X-coordinate must be first
   column in M
6. >>M(:,2) = W(:,8)   'Y-coordinate must be 2nd column
   in M
7. >>M(:,3) = W(:,1)   'Data, Z (in this case, visible
   blue reflectance) is third column
8. >>cd [CD-ROM drive letter]:\\MATLAB\\
9. >>pkrige_2(M, 5, 111., 482., 240., 1, 100, 100, 1.,
   -1., 1., 1., 'c:\\MATLAB\\NvLand6.grd')

or which the following syntax applies:

```
    pkrige_2([data: x, y, z], N, nugget, (sill-nugget),
range, model, nrows, ncols, sizex, sizey, ymax, xmin,
[filename]).
```

nce step 9 is executed, MATLAB prints results to the file that is specified in
ie call to pkrige_2. In the example shown in step 9, results are printed to a
le on the C drive, in a folder called matlab.

    This file is printed in a format that is compatible with *Visual_Data's* tool
or converting kriging results to a Windows bitmap image. Once the MATLAB
outine, pkrige_2, is completed, start *Visual_Data*, then open the MATLAB out-
ut file (e.g., NvLand6.grd) in the main window, and add the following line
 the beginning of the file:

```
NROWS    NCOLS    NUMDAT    IDAT    IY    IX
```

or which NROWS is the total number of rows in the grid, NCOLS is the total
umber of columns in the grid, NUMDAT is the total number of values on
ach line of the file, IDAT is the position on each line of the value to be used
 create the bitmap image, IY is the row number, and IX is the column num-
er. In the application to the *Nevada_Landsat_6x* data, NROW = NCOL =
J0; NUMDAT = 6; IDAT = 5; IY = 1, and IX = 2. The resultant bitmap
nage for the MATLAB results:

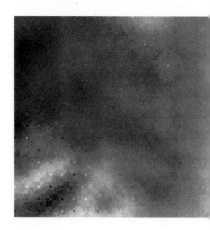

The MATLAB result is shown on the left, compared to the result from *Visu-al_Data* on the right. At first, a problem was suspected with the *Visual_Data* program. But, upon reinspection of the MATLAB routine, pkrige (Middleton 2000), it was discovered that the spherical variogram model was not imple-mented correctly:

```
Gam(i,j) = c * 1.5 * h - 0.5 * h^3
```

should have been coded as

```
Gam(i,j) = co + c (1.5 * h-0.5 * h^3).
```

This is now corrected in pkrige_2. The MATLAB result with the incorrect form of the spherical model, however, is a visually compelling result. Upon further reflection, it was determined that the incorrect form of the spherical model was, by accident, a linear model of the form, $\gamma(h) = mh$, where m in this case is 1.5c. A linear model is a valid, negative semi-definite variogram model. If $Gam(i,j) = c * 1.5 * h - 0.5 * h^3$ is used as a model, the term $0.5 * h^3$, is inconsequential compared to the term, $c * 1.5 * h$. That this is true is perhaps more obvious if it is explained that in the MATLAB routine, pkrige, $h = h/R$; that is, the distance of separation, h, is divided by R, the range. In this equation, then, h is always less than or equal to 1. Cubing this value and multiplying by 0.5 renders it even smaller. For all practical purposes, this term is equal to zero, and the spherical model, as coded, is accidentally a linear model.

     This is tested in *Visual_Data* by setting the following variogram parameters:

```
    1
    4   0.  173520.  240.  0.  1.
    1 100000.  5
```

Visual_Data does allow the use of linear model, model option = 4. A linear model never attains a sill. Yet, we still model a linear variogram using the stan-dard parameters, nugget, sill, range, and anisotropy. Internally, *Visual_Data* computes the slope, m, of the linear variogram model as: (Sill − nugget)/ Range. Thus, to match the MATLAB routine, pkrige, which inadvertently set $m = 1.5c$, and left the nugget out of the equation, we then set the nugget in *Vi-sual_Data* equal to zero, and set the sill = 1.5(482)(240) = 173520. The visual result from this application is shown for comparison:

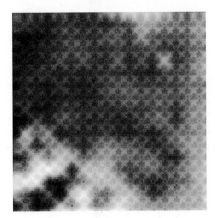

This result has an appearance more similar to the MATLAB outcome than that originally obtained from *Visual_Data* using a nugget value equal to 111. But, an artifact of kriging, a nonrandom pattern, is noticed in this outcome that was not seen in the MATLAB result. This prompted further investigation. The first hunch was that *Visual_Data* was executed again using the linear variogram model, but using a quadrant search algorithm to balance the nearest neighboring locations around each estimation location. This yielded the following visual result:

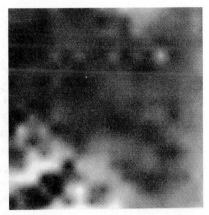

Left image: from *Visual_Data*, quadrant search. Center image: from MATLAB, pkrige_2 routine. Right image: from *Visual_Data*, general search, no correction for negative kriging weights.

The *Visual_Data* outcome is shown on the left in comparison to the MATLAB result shown in the center. The quadrant search strategy matches the MATLAB result closely. There are still some differences, most notably that the MATLAB result is smoother. Nevertheless, search strategy is seen to noticeably effect the estimation outcome.

This outcome suggests that *Visual_Data* has a bug when using the general search option. Of course, this was the suspicion. There is, however, no bug that is responsible for the artifact noticeable in the *Visual_Data* result using the general search strategy. Instead, it is acknowledged that *Visual_Data* sets negative kriging weights equal to zero. Negative weights are theoretically impossible in kriging. In practice, though, they do occur. In fact, negative weights are more likely when a nugget value of zero is used. They occur when more distant points are shadowed (screened) by closer points to the estimation location. The quadrant search strategy, by coincidence, resulted in less data shadowing thereby removing negative weights. By setting negative weights equal to zero a striping artifact was created in the estimation result because the negative weights were reflecting the very regular sampling geometry represented by these data, and by setting negative weights to zero, estimated values did not transition as smoothly as they would if these weights were left alone.

In comparison, the MATLAB routine, pkrige_2, does not alter negative kriging weights. More significantly, the visual outcome from this *MATLAB* routine suggests that negative kriging weights are quite important to the total estimation outcome and aid the smooth transition from one estimated value to the next. Truncating these negative weights by setting them equal to zero, as was done by *Visual_Data,* destroys this smooth transition and instead imparts a significant visual artifact. *Visual_Data* consequently was modified to remove the portion of the code that alters the value of negative kriging weights. Instead, an analyst can control the influence of negative weights by adjusting variogram parameters, especially the value of the nugget. Increasing the nugget value will decrease the numerical magnitude of negative weights, ultimately rendering all weights positive. More discussion on negative kriging weights and the screening effect is presented in Chiles and Delfiner (1999, pp. 204–205).

This discussion shows the value of applying two or more different computer algorithms in an analysis, thus allowing results to be compared. It is noted in closing this section that the MATLAB routine, pkrige_2, took eight hours to complete a 100 x 100 grid, whereas *Visual_Data* took 20 to 30 minutes for the same task.

5.10   Literature

Davis (1986) was inspirational for the presentation on time series analysis. The algebraic approach to describing the solution for kriging weights, and the discussion on self-affine fractal character of data, are both taken from Carr (1995). A notable, comprehensive, and recent book on geostatistics (spatial analysis) is that by Chiles and Delfiner (1999), which inspired many of the ideas for visual presentations of kriging results presented in this chapter. Throughout this text, Middleton (2000) was the important resource for MATLAB applications. In the case of the MATLAB kriging program, the MATLAB 5.3 Student Version manual was an important resource for understanding how to write output to a file.

A good companion to the present work is Isaaks and Srivastava (1989). Therein, a data set, topographic data from a location near Walker Lake, Nevada, is analyzed throughout the text to illustrate concepts. In the present work, Landsat data are used in a similar fashion. Furthermore, these Landsat data cover a region of western Nevada adjacent to Walker Lake, Mineral County, Nevada.

Exercises

1. The folder, \data\timedata\, on the CD-ROM contains the following sets of data:
   a. Sunspot.dat,
   b. Variable_Star.dat,
   c. Earthquake_Severity.dat,

Analyze these data sets using Fourier analysis to determine if any statistically significant cycles are present. If so, determine the time interval (wavelength) of each cycle. Then, comment on the significance of this time interval for each of these data sets.

2. The folder, \data\uscitym\, on the CD-ROM contains four temperature records for locations across the contiguous United States. The four records represent average high and average low temperature for January (1MN and 1MX) and July (7MN and 7MX). Arbitrarily select one temperature record for five different geographic locations and analyze each for statistically significant cycles. Comment on your findings.

3. Analyze the visible green, visible red, and near infrared reflectances (second, third, and fourth variables) of the *Nevada_Landsat_data* in the same manner as is shown in this chapter for visible blue reflectance:
   a. Compute omnidirectional variograms;
   b. Compute directional variograms;
   c. Apply kriging to obtain bitmap images of the 2x, 4x, and 6x resampled data.
      Comment on the characteristics of these variograms compared to those for the visible blue reflectance. What might you infer from differences? How do the bitmap images compare to those obtained for the visible blue reflectance?

4. Given: nugget, co, = 0. Sill = 1.0. R = 2.0. Using a spherical variogram model, show that kriging weights are negative for the more distant points in the geometric configuration shown below. Only examine weights. Data values are not shown at these data locations to minimize clutter in the diagram. What happens to these negative weights if a nugget value of 0.5 is used? Diagram:

5. A linear model yielded particularly good visual results in application to the *Nevada_Landsat_6x* data set. This was a serendipitous discovery owing to the incorrect coding of the spherical variogram model in the MATLAB routine, pkrige (and pkrige_2). Nonetheless, the application of this linear model shows the visual quality of using a variogram model that doesn't match the actual data variogram. Show mathematically why this model renders a visually superior result to that suggested by the data varuogram.

6. Use the linear variogram model to form a bitmap image of the *Nevada_Landsat_4x* data set and compare to the result shown earlier in this chapter.

7. Using the data configuration shown in Section 5.2.1 that was used to demonstrate the calculations involved in variogram analysis:
   a. continue this demonstration by computing the variogram values for $h = 3$ and $h = 4$.
   b. by hand, transform the data values to indicator values using $c = 75$. Then, compute by hand variogram values for $h = 1, h = 2, h = 3$, and $h = 4$. How do these values compare to those for the raw data in terms of the growth of the variogram from the origin?

8. Analyze the visible green, red, and near infrared variables (variables 2, 3, and 4) of the *Nevada_Landsat_data* using indicator transform. For each, set the cut off, c, equal to the median. Develop bitmap images of the results, and comment on the spatial location of higher and lower data values.

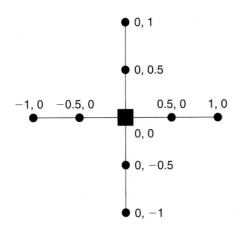

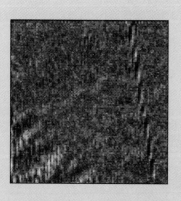

6

Multivariate Spatial Data Analysis

Now that kriging has been introduced for the estimation of a single variable distributed in space, an extension to the estimation of multiple variables distributed in the same space is described. This involves the generalization of kriging to become *cokriging*. Synonymous with estimation, kriging refers to the more simple univariate case, whereas cokriging implies the estimation of more than one quantity. Spatial autocovariance is the fundamental information on which estimation is based. But, the relationship between variables is also used. Cokriging is consequently based on more information in comparison to kriging, and theoretically is associated with smaller estimation variance compared to kriging the multiple variables separately.

Throughout this text, the primary data set that has been analyzed, thus providing a unifying theme, is the *Nevada_Landsat* data set. These data represent seven variables distributed within the same space. In the previous chapter, the focus was solely on one of these variables, visible blue reflectance. In the present chapter, cokriging is used to estimate all seven variables simultaneously. As always, visualizations are presented to demonstrate the properties inherent to cokriging, a *multivariate, geostatistical estimator*.

6.1 Theory

Recall the formula for the univariate case (kriging):

$$Z^*(x_0) = \sum_{i=1}^{N} \lambda_i Z(x_i); \quad \sum_{i=1}^{N} \lambda_i = 1$$

[a scalar estimate is equal to a weighted combination of *N* closest known scalar values]

Using this same concept, the cokriging estimator is written as

$$\overline{Z^*}(x_0) = \sum_{i=1}^{N} T_i \overline{Z}(x_i); \quad \sum_{i=1}^{N} T_i = I$$

[a vector estimate is equal to a weighted combination of *N* closest known vectors]

In this formula, the overline symbol above Z signifies its vector representation; in this case, Z is a multivariate vector, $(Z_1(x), Z_2(x), \dots, Z_M(x))$, M variables in size. There is no theoretical limit to M. Further in this equation, T is a **matrix** of weights, $M \times M$ in size. Collectively, these N weighting matrices sum to the identity matrix, I, also $M \times M$ in size.

Suppose $M = 2$, and further suppose that N, the number of nearest neighboring data locations used in cokriging, is also 2. Then, we have the following:

$$\begin{vmatrix} Z_1^*(x_0) \\ Z_2^*(x_0) \end{vmatrix} = |T_1| \begin{vmatrix} Z_1(x_1) \\ Z_2(x_1) \end{vmatrix} + |T_2| \begin{vmatrix} Z_1(x_2) \\ Z_2(x_2) \end{vmatrix}$$

This is expanded as follows to make the weighting matrices, T, more obvious:

$$\begin{vmatrix} Z_1^*(x_0) \\ Z_2^*(x_0) \end{vmatrix} = \begin{vmatrix} \lambda_{11,1} & \rho_{12,1} \\ \rho_{21,1} & \lambda_{22,1} \end{vmatrix} \begin{vmatrix} Z_1(x_1) \\ Z_2(x_1) \end{vmatrix} + \begin{vmatrix} \lambda_{11,2} & \rho_{12,2} \\ \rho_{21,2} & \lambda_{22,2} \end{vmatrix} \begin{vmatrix} Z_1(x_2) \\ Z_2(x_2) \end{vmatrix}$$

Carrying out the matrix algebra helps to show how cokriging works:

$$Z_1^*(x_0) = \lambda_{11,1}Z_1(x_1) + \rho_{12,1}Z_2(x_1) + \lambda_{11,2}Z_1(x_2) + \rho_{12,2}Z_2(x_2)$$
$$Z_2^*(x_0) = \rho_{21,1}Z_1(x_1) + \lambda_{22,1}Z_2(x_1) + \rho_{21,2}Z_1(x_2) + \lambda_{22,2}Z_2(x_2)$$

Notice that the weights, λ, are applied to the same variable that is being estimated. These are the primary weights. Further, the weights, ρ, are applied to the variable(s) other than the one that is being estimated. These are the secondary weights. Further, given that $\Sigma T = I$, then the following is implied:

$$\sum_{i=1}^N \lambda_{jj,i} = 1; \quad \sum_{i=1}^N \rho_{jk,i} = 0; j = 1, 2, \ldots, M; k = 1, 2, \ldots, M; k \neq j.$$

Because the secondary weights sum to zero, cokriging is an unbiased estimator. Moreover, the estimate of each variable is a function of *all* variables. This distinguishes cokriging from kriging.

6.1.1 Solving for the Cokriging Weights

In Chapter 5, it is shown that kriging weights are a function of spatial autocovariance, captured by the variogram. The Lagrangian expression used in Chapter 5 to obtain a solution for kriging weights is also the expression that is used to obtain a solution for cokriging weights, except scalar values are replaced by matrices and vectors. The variogram is still necessary to the solution of cokriging weights. Additionally, a function known as the **cross-variogram** is necessary for representing the cross-spatial autocovariance between two variables.

6.1.1.1 The Matrix System: Kriging

In ordinary, punctual kriging, the matrix system used to compute weights is:

$$\begin{vmatrix} \lambda_1 \\ \lambda_2 \\ \cdot \\ \cdot \\ \lambda_N \\ \mu \end{vmatrix} = \begin{vmatrix} Cov(x_1x_1) & Cov(x_1x_2) & \ldots & Cov(x_1x_N) & 1 \\ Cov(x_2x_1) & Cov(x_2x_2) & \ldots & Cov(x_2x_N) & 1 \\ \cdot & \cdot & \ldots & \cdot & \cdot \\ \cdot & \cdot & \ldots & \cdot & \cdot \\ Cov(x_Nx_1) & Cov(x_Nx_2) & \ldots & Cov(x_Nx_N) & 1 \\ 1 & 1 & \ldots & 1 & 0 \end{vmatrix}^{-1} \begin{vmatrix} Cov(x_0x_1) \\ Cov(x_0x_2) \\ \cdot \\ \cdot \\ Cov(x_0x_N) \\ 1 \end{vmatrix}$$

for which Gauss elimination is usually used to perform a pseudo-inverse of the covariance matrix. The covariance entries are obtained from the variogram as $Cov(h) = Sill - \gamma(h)$, moreover h, separation distance, is the Euclidean distance between two spatial locations, x_i and x_j.

6.1.1.2   The Matrix System: Cokriging:

In ordinary, punctual cokriging, the matrix system used to compute weights is

$$
\begin{vmatrix} T_1 \\ T_2 \\ \cdot \\ \cdot \\ T_N \\ \overline{\mu} \end{vmatrix} = \begin{vmatrix} \overline{Cov}(x_1x_1) & \overline{Cov}(x_1x_2) & \dots & \overline{Cov}(x_1x_N) & I \\ \overline{Cov}(x_2x_1) & \overline{Cov}(x_2x_2) & \dots & \overline{Cov}(x_2x_N) & I \\ \cdot & \cdot & \dots & \cdot & \cdot \\ \cdot & \cdot & \dots & \cdot & \cdot \\ \overline{Cov}(x_Nx_1) & \overline{Cov}(x_Nx_2) & \dots & \overline{Cov}(x_Nx_N) & I \\ I & I & \dots & I & 0 \end{vmatrix}^{-1} \begin{vmatrix} \overline{Cov}(x_0x_1) \\ \overline{Cov}(x_0x_2) \\ \cdot \\ \cdot \\ \overline{Cov}(x_0x_N) \\ I \end{vmatrix}
$$

in which each entry is actually an $M \times M$ matrix.

This is looking very complicated! As is the case with all chapters in this text, substantive insight to theory is presented to enable readers to understand how the computer software used in data analysis operates. Equations are solved by the software, not the analyst. But, the analyst feeds the software information, necessitating an understanding of how this information is manipulated to enable an assessment of the adequacy of the outcome. Computer software should not be used as a "black box" because an analyst, or researcher, or student, has no idea what the software is doing to the data.

Returning to the matrix system for cokriging, the primary reason this matrix system is presented is to show where and why the cross-variogram is necessary. Focusing on one of the covariance entries in the matrix system:

$$
\overline{Cov}(x_ix_j) = \begin{vmatrix} Cov_{ij,11} & CCov_{ij,12} & \dots & CCov_{ij,1M} \\ CCov_{ij,21} & Cov_{ij,22} & \dots & CCov_{ij,2M} \\ \cdot & \cdot & \dots & \cdot \\ \cdot & \cdot & \dots & \cdot \\ CCov_{ij,M1} & CCov_{ij,M2} & \dots & Cov_{ij,MM} \end{vmatrix}
$$

This formula is generic and holds for any i and j, including $i = 0$. Notice the diagonal entries are written as $Cov_{ij, kk}$. Each of these entries is computed using the variogram for variable k, and k ranges from 1 to M, the number of variables being considered in cokriging. Consequently, we see the need to model M variograms when performing cokriging.

Additional entries in this matrix are written as $CCov_{ij, ab}$. This represents the **cross-covariance** between variables, a and b, at locations, i and j. Cross-covariance is computed as follows:

$$
CCov_{ij,ab} = \frac{1}{2}[Cov_{ab}^+(h_{ij}) - Cov_a(h_{ij}) - Cov_b(h_{ij})]
$$

[cross-covariance is equal to one-half the sum of the covariance of the paired sum, $a + b$, the negative of the covariance on variable a, alone, and the negative of the covariance of variable b, alone]

This introduces a new notion, that of the **paired-sum,** $a + b$. To compute cross-covariance between variables, a and b, their sum, $a + b$, is computed at each and every spatial location at which both a and b are sampled. A variogram is then computed for this "new" variable, the paired-sum, and modeled using one of the valid variogram models, such as spherical, or exponential. This variogram is designated as $\gamma^+(h)$. From this is computed the covariance for the paired sum, $Cov^+(h) = Sill^+ - \gamma^+(h)$. By this representation, we see that the cross-covariance, $CCov$, is obtained as a linear combination of covariance values, each a function of a variogram.

In cokriging, a paired-sum variogram is required for each two-variable combination, ab, possible from the collection of M variables being estimated. If M is 2, there is only one possible combination: that for variables 1 and 2. If M is 3, there are three possible pairs: variables 1 and 2, variables 1 and 3, and variables 2 and 3. In general, there are $M(M - 1)/2$ possible combinations, and this many paired-sum variograms must be modeled.

DEMONSTRATION

Nevada_Landsat_6x data set. There are seven variables associated with this data set: visible blue, visible green, visible red, near infrared, mid-infrared, thermal infrared, and mid-infrared. Eventually, these seven variables will be treated using cokriging. In this case, 7 variograms and $7(7 - 1)/2$, or 21 paired-sum variograms will need to be modeled, for a total of 28 variogram models!

Visual_Data is used to compute the paired-sum variograms (also called cross-variograms). When using its cross-variogram tool, *Visual_Data* first computes variograms for the M variables, then computes paired-sum variograms for the $M(M - 1)/2$ paired combinations. Only a few visual examples are shown, and a table summarizes the chosen models for all 7 variograms and 21 paired-sum variograms.

Visual_Data provides an option to rescale each of the M variables to a numeric range, 0 to 1. Rescaling is quite important to cokriging for several reasons. If cross-covariance is a function of paired-sum variograms, then a large numeric magnitude differential between the two variables will result in a paired-sum variogram that is tantamount to the variogram for the variable having the larger numeric magnitude. Rescaling corrects this possibility. During estimation (cokriging), rescaling can result in improved estimation, again because data magnitude differential is corrected, thus allowing the M variables to influence estimation performance more equally. Data rescaling was used to obtain the variograms and paired-sum variogram shown in Figures 6.1–6.3.

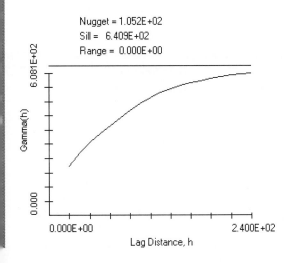

Nugget = 1.052E+02
Sill = 6.409E+02
Range = 0.000E+00

◀ FIGURE 6.1
Variogram for the visible red variable. Morphologically, this variogram is similar to that for visible blue reflectance that is presented in Chapter 5.

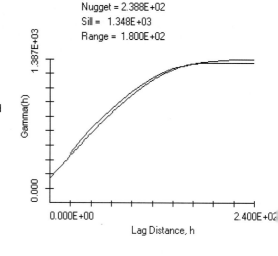

▶ **FIGURE 6.2**
Variogram for the mid-infrared reflectance, variable 5. The data variogram is modeled quite closely as a spherical structure. The sill is attained and is exactly equal to sample variance. This structure is distinctly different from what is observed for visible blue and red reflectances.

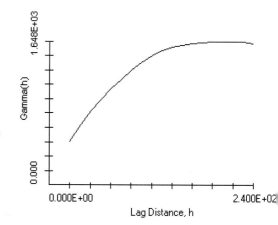

▶ **FIGURE 6.3**
Paired-sum variogram for the combination of variable 6 (thermal infrared) and variable 7 (mid-infrared). This result is not a model, but is the actual data outcome displaying perfectly spherical variogram structure for the paired-sum: thermal+band 7.

　　　Model parameters for all 28 variograms are shown in the tables below. Cokriging is seen to involve substantially more work when modeling variograms in comparison to kriging. We could approach this problem as seven separate krigings. In this case, only seven variograms are needed. However, additional infor-

Variogram parameters, *Nevada_Landsat_6x* data set.						
Variable	**Model**	**Nugget**	**Sill**	**Range**	**Anisotropy**	**Ratio**
1: vis. blue	Exponential	0.004	0.020	180.	0.	1.
2: vis. green	Exponential	0.004	0.018	180.	0.	1.
3: vis. red	Exponential	0.004	0.020	180.	0.	1.
4: near IR	Spherical	0.003	0.012	240.	0.	1.
5: mid-IR	Spherical	0.004	0.022	180.	0.	1.
6: thermal	Spherical	0.005	0.037	180.	0.	1.
7: mid-IR	Spherical	0.005	0.028	156.	0.	1.

Paired-Sum Variograms, *Nevada_Landsat_6x* data set.						
Pair	Model	Nugget	Sill	Range	Anisotropy	Ratio
1 and 2	Exponential	0.015	0.074	180.	0.	1.
1 and 3	Exponential	0.015	0.077	180	0.	1.
1 and 4	Exponential	0.008	0.049	180.	0.	1.
1 and 5	Spherical	0.015	0.075	200.	0.	1.
1 and 6	Spherical	0.01	0.093	192.	0.	1.
1 and 7	Spherical	0.02	0.092	216.	0.	1.
2 and 3	Exponential	0.015	0.076	220.	0.	1.
2 and 4	Spherical	0.008	0.049	180.	0.	1.
2 and 5	Spherical	0.015	0.074	200.	0.	1.
2 and 6	Spherical	0.009	0.090	192.	0.	1.
2 and 7	Spherical	0.009	0.090	220.	0.	1.
3 and 4	Spherical	0.010	0.054	240.	0.	1.
3 and 5	Spherical	0.016	0.081	200.	0.	1.
3 and 6	Spherical	0.009	0.095	192.	0.	1.
3 and 7	Spherical	0.020	0.097	220.	0.	1.
4 and 5	Spherical	0.015	0.061	200.	0.	1.
4 and 6	Spherical	0.007	0.070	180.	0.	1.
4 and 7	Spherical	0.014	0.070	200.	0.	1.
5 and 6	Spherical	0.009	0.095	204.	0.	1.
5 and 7	Spherical	0.020	0.100	216.	0.	1.
6 and 7	Spherical	0.011	0.110	204.	0.	1.

mation describing the spatial intercorrelation among variables cannot be included. The experiment is to determine if the additional work involved with cokriging is worth the trouble.

The same grid that was used in Chapter 5 is used also used in this experiment, primarily to enable comparisons between kriging and cokriging. Seven bitmap images, one for each of the seven variables, is obtained in this experiment, displayed below with the original images for visual comparison.

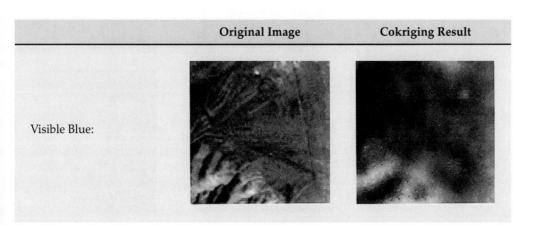

	Original Image	Cokriging Result
Visible Blue:		

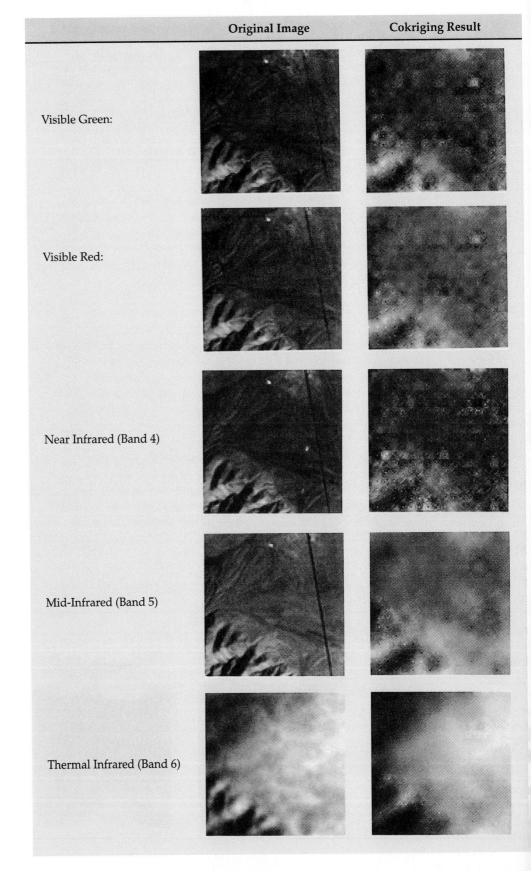

Original Image	Cokriging Result

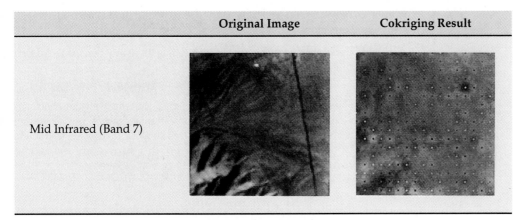

Mid Infrared (Band 7)

6.2   On the Practice of Cokriging

The foregoing demonstration illustrates some interesting aspects of spatial coregionalization.

Recall from Chapter 3 that the statistical correlation among these seven variables is complex and often nonlinear. In cokriging, the cross-variogram is not related to statistical correlation. Instead, it reflects how two variables vary with respect to each other *within the same space.* This is a different notion than statistical correlation. Cokriging can proceed in the absence of statistical correlation between two variables, provided a useful cross-variogram is found.

Data rescaling was introduced in the foregoing demonstration to achieve parity in numerical magnitude across variables. This is an essential aspect of cokriging, otherwise larger magnitude variables will disproportionately influence estimation outcome. The rescaling is simply

```
Z'(x) = (Z(x) - Minimum)/(Maximum - Minimum)
```

in which **Minimum** is the smallest data value (or the most negative) and **Maximum** is the largest data value. The rescaling is conducted for each variable separately, not across variables. In other words, Minimum and Maximum are identified for each variable, then rescaling is completed, and this is repeated for the next variable, and so on.

Variograms and paired-sum variograms are computed using the rescaled values. Estimation also proceeds using the rescaled values. A final step after cokriging estimation rescales the estimated values back to the original numerical range:

```
Z*(x) = ((Maximum - Minimum) * Z*'(x)) + Minimum
```

in which $Z^{*\prime}(x)$ is the cokriging estimate of one of the variables, Z, at a location, x, based on rescaled values, and $Z^*(x)$ is the estimated value in the context of original data magnitude.

DISCUSSION

Empirical Insights to Cokriging Through Cross-Validation

Cross-validation was introduced in Chapter 5 as a "leaving-one-out" procedure for testing the performance of kriging. This method is equally useful for examining the performance of cokriging. In this discussion, cross-validation is used to investigate the following questions:

1. If the M variables are highly intercorrelated, is there any benefit to cokriging over kriging?
2. Is data rescaling really necessary for cokriging?

We begin with the first question. The *Nevada_Landsat_data* comprise seven variables that exhibit a high degree of intercorrelation (Chapter 3). Cokriging has already been applied to these data earlier in this chapter. It remains to document the accuracy of this application. For the moment, this discussion focuses on estimation performance for visible reflectance, blue, green, and red.

Cokriging yields the same accuracy as does kriging for visible blue reflectance

Cokriging:

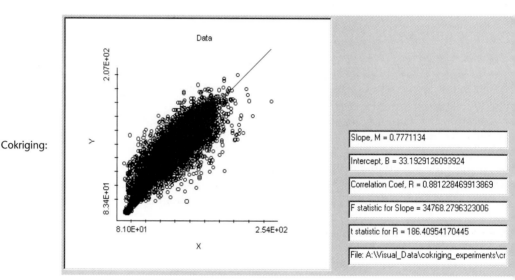

Kriging:

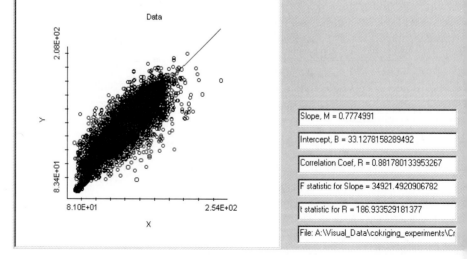

Moreover, cokriging and kriging yield comparable accuracy in application to visible green reflectance:

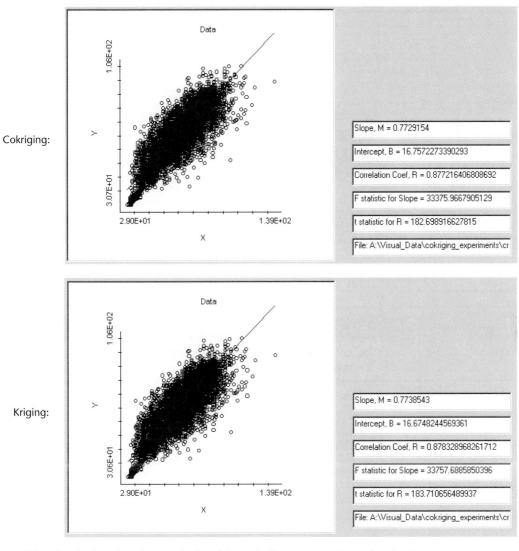

Cokriging:

Slope, M = 0.7729154

Intercept, B = 16.7572273390293

Correlation Coef, R = 0.877216406808692

F statistic for Slope = 33375.9667905129

t statistic for R = 182.698916627815

File: A:\Visual_Data\cokriging_experiments\cr

Kriging:

Slope, M = 0.7738543

Intercept, B = 16.6748244569361

Correlation Coef, R = 0.878328968261712

F statistic for Slope = 33757.6885850396

t statistic for R = 183.710656489937

File: A:\Visual_Data\cokriging_experiments\cr

Likewise, both estimation methods achieve similar accuracy in application to the estimation of visible red reflectance:

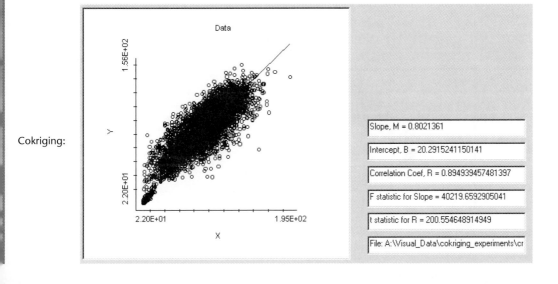

Cokriging:

Slope, M = 0.8021361

Intercept, B = 20.2915241150141

Correlation Coef, R = 0.894939457481397

F statistic for Slope = 40219.6592905041

t statistic for R = 200.554648914949

File: A:\Visual_Data\cokriging_experiments\cr

Kriging:

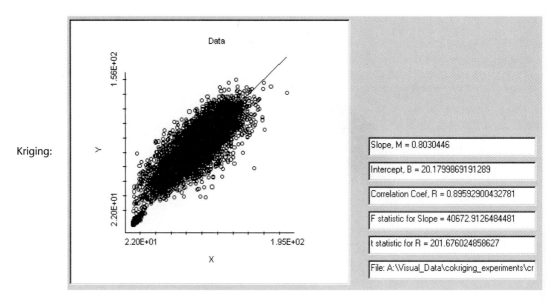

There are subtle differences in these results, but they are not so significant that one estimation approach can be said to be "better" than another. In one additional experiment, cross-validation is used to examine the estimation performance of cokriging using only $M = 2$ variables, near infrared (band 4) and thermal (band 6). In Chapter 3, these two variables among those of which the *Nevada_Landsat_data* comprise are found to represent the poorest statistical correlation. Consequently, they represent the best test for these data of the effect of statistical correlation on cokriging performance.

The results for the estimation of thermal emission are almost identical, but that from cokriging is slightly better judging by the *F*-statistic for regression performance. Incidentally, the method of regression that should be used when analyzing results from cross-validation is that which presumes all error is on y (the estimated values).

Kriging:

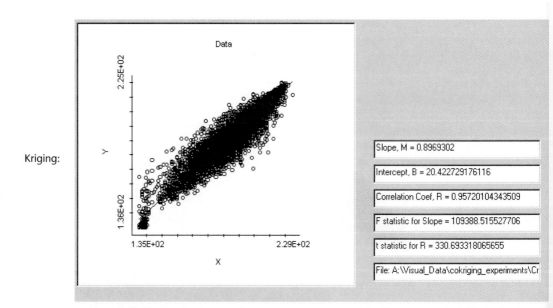

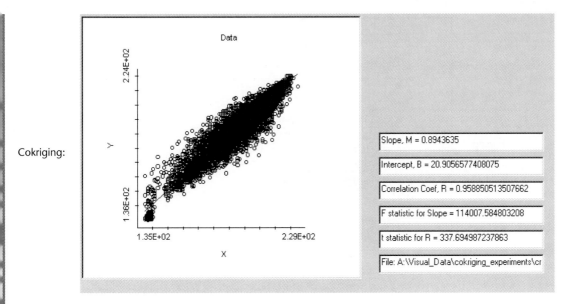

Cokriging:

In these examples, kriging and cokriging are fairly indistinguishable with respect to estimation performance. Only when statistical correlation is weaker (near infrared with thermal) was any distinction possible between these estimation methods. From an empirical stand point, these results suggest that no improvement in estimation performance is realized when using cokriging instead of kriging for these Landsat data.

Application to a single data set is not sufficiently exhaustive to constitute proof of estimation performance. A further test is presented using a 1986 image of Brazilian rainforest, acquired by the French SPOT satellite. This satellite system acquires multispectral data in three electromagnetic bands: visible green, visible red, and near infrared. Cokriging is applied to the estimation of all three variables in cross-validation mode. Choosing near infrared as a test, mostly because this is the dominant reflectance observable in this scene given the dense vegetation of the rainforest, yields the following results from cokriging and kriging:

Cokriging:

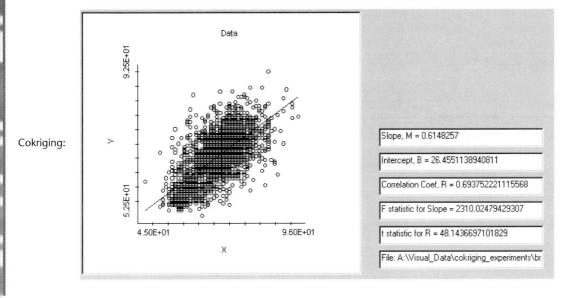

Kriging:

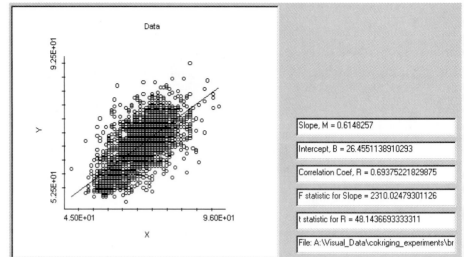

Again, no difference is discernable in these outcomes. Correlation coefficients among these spectral bands are presented in the table below (based on regression assuming equal error on x and y):

Spectral Band	Visible Green	Visible Red	Near Infrared
Visible Green	1.0	0.96 (linear)	−0.5 (non-linear)
Visible Red		1.0	−0.6 (non-linear)
Near Infrared			1.0

Correlation among these variables is not particularly strong. Even so, parity exists between kriging and cokriging in terms of estimation performance. Although not shown, further experiments were attempted that involved different variogram and cross-variogram parameters. In particular, the nugget value for all three variogram and three cross-variogram models was set to zero. Manipulation of these parameters had no consequence for the cokriging outcome. Kriging and cokriging results remained identical.

A further test is conducted using a data set, Jura.dat, listed in Goovaerts (1997), and found on the CD in the directory, \data\othrdata\. These data comprise seven variables: cadmium, copper, lead, cobalt, chromium, nickel, and zinc. Correlation coefficients among these variables are listed in the following table:

	cadmium	copper	lead	cobalt	chromium	nickel	zinc
cadmium	1.0	0.12	0.22	0.25	0.61	0.49	0.67
copper		1.0	0.78	0.22	0.21	0.23	0.57
lead			1.0	0.19	0.29	0.31	0.59
cobalt				1.0	0.45	0.75	0.47
chromium					1.0	0.69	0.67
nickel						1.0	0.63
zinc							1.0

Lesser statistical correlation is noticed among these variables in comparison to the *Nevada_Landsat_data*. An experiment is conducted to compare the performance of cokriging to that of kriging in application to these data. For the sake of brevity, visual examples focus on variables 1 (cadmium) and 4 (cobalt):

Variable 1:
Cokriging:

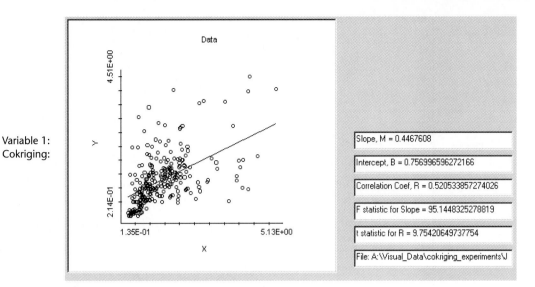

Slope, M = 0.4467608

Intercept, B = 0.756996596272166

Correlation Coef, R = 0.520533857274026

F statistic for Slope = 95.1448325278819

t statistic for R = 9.75420649737754

File: A:\Visual_Data\cokriging_experiments\J

Kriging:

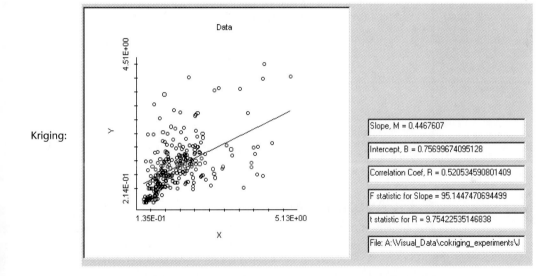

Slope, M = 0.4467607

Intercept, B = 0.75699674095128

Correlation Coef, R = 0.520534590801409

F statistic for Slope = 95.1447470694499

t statistic for R = 9.75422535146838

File: A:\Visual_Data\cokriging_experiments\J

Variable 4:
Cokriging:

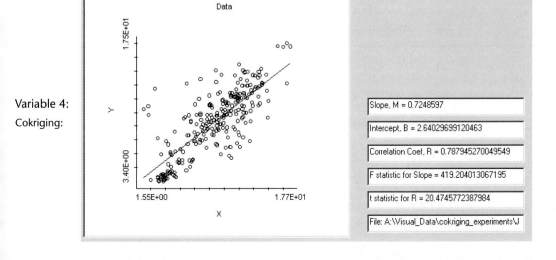

Slope, M = 0.7248597

Intercept, B = 2.64029699120463

Correlation Coef, R = 0.787945270049549

F statistic for Slope = 419.204013067195

t statistic for R = 20.4745772387984

File: A:\Visual_Data\cokriging_experiments\J

Kriging:

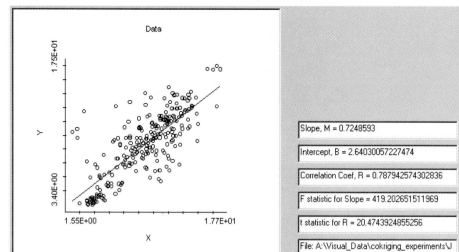

These results are based on the following variogram and cross-variogram parameters (for rescaled parameters):

Variable 1:	Nugget = 0	Sill = 0.013	Range = 1.0 (km) Spherical
Variable 4:	Nugget = 0	Sill = 0.049	Range = 1.0 (km) Spherical
Paired Sum:	Nugget = 0	Sill = 0.146	Range = 1.0 (km) Spherical

No improvement in estimation performance is observed when applying cokriging to these data. Estimation performance seems to be insensitive to statistical correlation among the multiple variables.

6.3   Autokrigeability

Foregoing examples seem to suggest that there is no advantage to cokriging over kriging. The variables treated in these examples appear to be **autokrigeable** (e.g., Wackernagel, 1995). A variable is said to be autokrigeable if kriging and cokriging yield equivalent results. Variables that are statistically independent are autokrigeable because only the variable being estimated will receive nonzero weighting, whereas the weights on all other variables are zero. Additionally, variables that exhibit intrinsic correlation are autokrigeable.

Intrinsic correlation is said to pertain to multivariate data whose multivariate correlation is independent of spatial correlation. Such is the case with satellite data. Multivariate correlation is a function of variable spectral response of terrain features, regardless of where these features occur. Several examples are presented to explore the notion of autokrigeability.

6.3.1   Application to Principal Components Images

Principal components transformation was introduced in Chapter 4. One way to test the autokrigeability of a set of spatial data is to apply principal components transformation and compare the performances of kriging and cokriging in application to the estimation of the principal components (Wackernagel, 1995, p. 142).Principal components are computed of the *Nevada_Landsat_data* in Chapter 4. Applying kriging and cokriging to these data yields the following results:

Cokriging: (1st PC image):

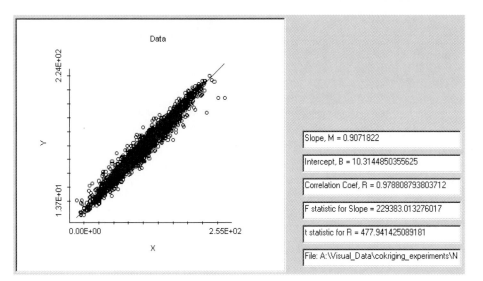

Kriging: (1st PC image)

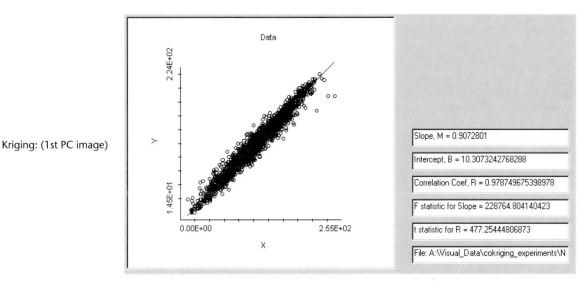

The method of standardized principal components analysis is used in this example, yielding results from kriging and cokriging that are identical. This represents empirical support for concluding that intrinsic correlation is a valid assumption in the case of the *Nevada_Landsat_data*. This realization is consistent with results obtained earlier in this chapter showing similar, or identical results when applying kriging and cokriging to the raw data. When an intrinsic correlation exists, kriging variables separately is faster and yields the same results as would be obtained with cokriging.

6.3.2   Extension to Indicator Cokriging

The indicator transform was introduced in Chapter 5. In the present example, the indicator transform represents a method for demonstrating what happens when variables do not exhibit intrinsic correlation.

Visible blue reflectance is the chosen variable among those comprising the *Nevada_Landsat_6x* data set. Three indicator transforms are formed using cutoffs equal to the 25, 50 (median), and 75 percentiles. These variables are estimated simultaneously using cokriging based on the following variogram parameters:

Variable	Model	Nugget	Sill	Range	Anisotropy	Ratio
$i(x)$, 25%	Spherical	0.0	0.17	240.	0.	1.
$i(x)$, 50%	Spherical	0.0	0.24	240.	0.	1.
$i(x)$, 75%	Spherical	0.0	0.18	120.	0.	1.
25 + 50	Spherical	0.0	0.66	240.	0.	1.
25 + 75	Spherical	0.0	0.46	120.	0.	1.
50 + 75	Spherical	0.0	0.64	240.	0.	1.

Speculatively, these three indicator transforms should not exhibit intrinsic correlation because they are based on the same spatial variable, visible blue reflectance. It remains to test this notion empirically by comparing the results from cokriging to those from kriging. The median is chosen as the cutoff for comparison. Kriging is applied to the estimation of the indicator values formed using the median as the cutoff and the variogram parameters shown above for $i(x)$, 50%.

Bitmap images are formed from the cokriging and kriging estimation results:

Cokriging   Kriging

These are identical results, suggesting that these indicator variables are autokrigeable. Once again, an example is encountered in which no benefit is derived from the application of cokriging in alternative to kriging.

6.4   The Undersampled Case

Indeed, Wackernagel (1995, p. 144) explains:

> " . . . if all variables have been measured at all sample locations and if the variables are intrinsically correlated, then cokriging is equivalent to kriging."

All examples thus far presented in this chapter involve such fully sampled data. **Fully sampled data** are multivariate, spatial data for which all variables are sampled at all spatial locations. In this case, the sampling resolutions of the variables are equal. For such data, kriging is the preferred estimator because its accuracy matches that of cokriging without the need to model cross-variograms.

There is, nonetheless, a distinct advantage to using cokriging over kriging when the sampling resolutions of the multiple variables differ. In this case, some variables are not sampled at some locations. This is known as **undersampling.** In cokriging applied to undersampled, multivariate spatial data, the resolution of one or more variables is improved by benefitting from the variable, or variables, associated with the highest resolution.

Coarser sampled variables are deficient in higher spatial frequency components that are present in more finely sampled data. In cokriging, recall that weights, ρ, applied to auxiliary variables (those other than the one being estimated) sum to zero. It will be shown in Chapter 8 on Digital Image Processing that weighting pixels using values summing to zero yields a **high-pass filter** of these pixels. Such a filter enhances, or preserves (passes) high spatial frequencies, whereas lower frequencies are dampened or removed. Cokriging can be viewed as a method in which high frequency information is added to variables deficient in such information.

When data are fully sampled, the spatial frequency content is the same for all variables. No variable is deficient with respect to another variable in spectral content. Consequently, no improvement is possible when estimating a variable using its spatial covariance information in conjunction with cross-covariance information with other variables. This is another way of explaining the identical results from kriging and cokriging in application to fully sampled data that does not require the concept of intrinsic correlation.

6.4.1   Accommodating Undersampling in Cokriging

Fundamentally, undersampling is accommodated in cokriging by modifying the covariance matrix system:

$$
\begin{vmatrix} T_1 \\ T_2 \\ \cdot \\ \cdot \\ T_N \\ \overline{\mu} \end{vmatrix} =
\begin{vmatrix}
\overline{Cov}(x_1x_1) & \overline{Cov}(x_1x_2) & \ldots & \overline{Cov}(x_1x_N) & I \\
\overline{Cov}(x_2x_1) & \overline{Cov}(x_2x_2) & \ldots & \overline{Cov}(x_2x_N) & I \\
\cdot & \cdot & \ldots & \cdot & \cdot \\
\cdot & \cdot & \ldots & \cdot & \cdot \\
\overline{Cov}(x_Nx_1) & \overline{Cov}(x_Nx_2) & \ldots & \overline{Cov}(x_Nx_N) & I \\
I & I & \ldots & I & 0
\end{vmatrix}^{-1}
\begin{vmatrix}
\overline{Cov}(x_0x_1) \\
\overline{Cov}(x_0x_2) \\
\cdot \\
\cdot \\
\overline{Cov}(x_0x_N) \\
I
\end{vmatrix}
$$

Undersampled variables are represented by zeros in the data set. When encountering these zero values, the computer code is designed to "zero-out" the row position and column position in the covariance matrix system that corresponds to the missing value. In this event, the missing value is estimated based only on auxiliary information.

6.4.2   Application to the *Nevada_Landsat* Data Set: Improving the Resolution of Thermal Images

Seven spectral bands are scanned by the Landsat satellite. Six of these bands are sampled at 30 meter resolution: visible blue (band 1), visible green (band 2), visible red (band 3), near infrared (band 4), mid-infrared (band 5), and mid-infrared (band 7). The thermal band, band 6, is sampled at 120 m resolution, four times coarser than the optical bands. A different sensor is required to detect thermal **emission,** one that must be sensitive to heat, whereas the other sensors respond to **reflectance.** The differing sensor systems, in part, explains the differing resolutions.

An experiment is attempted to improve the spatial resolution of the thermal image using one of the other, higher resolution bands. In this case, band 7, mid-infrared, is arbitrarily chosen. Only one auxiliary band is used in this case, and cokriging will proceed based on just two variables. This implementation is chosen for two reasons, only one high resolution band is necessary to effect higher resolution in the thermal band, moreover a two-variable case executes more rapidly in the computer software domain than does the seven variable case.

Full resolution landsat data is used in this example (*Nevada_Landsat_1x* data set). The thermal image is shown earlier in this chapter and is blurry in comparison to other six bands. This is a consequence of its lower resolution. An experiment is conducted to improve the thermal resolution by taking the fully sampled, thermal data and setting every other pixel value equal to zero. Granted, this will improve its resolution only two-fold, not four-fold. But, zeroing all but every fourth pixel value will yield a much sparser data system, and this may cause numerical instability in cokriging. (There is a limit to how much undersampling cokriging can accommodate before its covariance matrix systems become so sparse that numerical errors begin to occur during equation solution).

Cokriging is applied to the subsampled thermal data and fully sampled band 7 data without using rescaling. Variogram and cross-variogram parameters for thermal, band 7, and the paired-sum, thermal + band 7, are those presented earlier in this chapter. (Note: those variogram parameters that are presented earlier are for rescaled data. Those variogram parameters are used here, even though rescaling is not used during estimation. What is of paramount importance to this example are the cokriging weights, not the cokriging variance. In this case, the rescaled variograms present a more correct picture of the actual spatial relationships because all variables are of equal magnitude).

To avoid confusion, the first five lines of the resampled data are compared to the fully sampled data to show how every other thermal value is set to zero. (Recall that a zero value indicates not sampled, or undersampled):

	6	7	Y	X
Fully Sampled:	181	90	1	1
	182	91	1	2
	182	93	1	3
	181	93	1	4
	182	95	1	5
Undersampled:	0	90	1	1
	182	91	1	2
	0	93	1	3
	181	93	1	4
	0	95	1	5

Visual results show the value of cokriging when undersampling is a characteristic of a set of data. In this case, the cokriging result is compared to the original image:

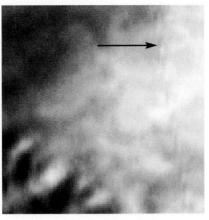

The original thermal band, left image, is compared to the cokriging result, right image. The arrow highlights a linear feature (road) that appears in the cokriging result that is not apparent in the original image.

The improvement from cokriging is perhaps not that obvious. One way to compare the difference between two images is by image subtraction. *Visual_ Data* provides a tool for digital image subtraction. In this case, the kriging result is subtracted from the cokriging result to show the difference in the two estimation methods.

Kriging   Cokriging   Difference

The difference image shows that the cokriging result does differ from the kriging result. The linear detail (road) is distinctly apparent in the difference image. This linear feature is clearly evident in six of the seven images shown earlier in this chapter; it is not apparent in the original thermal image either because its resolution is too coarse to reveal the road, or the thermal signature of the road matches exactly that of the surrounding terrain. In this application of undersampled cokriging, the higher resolution auxiliary information, band 7, enabled higher resolution features, such as roadways, to be resolved in the thermal image.

6.5   How Do I Reproduce Results in This Chapter Using . . .

6.5.1   *Visual_Data?*

As with other applications presented in this text, applying *Visual_Data* for cokriging begins by opening the primary data set in the main program window. Once *Visual_Data* is started from the Windows Program menu, a user clicks File, then Open, and browses to find the file. In the case of the

Nevada_Landsat_6x data, for instance, the data are found on the CD-ROM in the folder, \data\nvlandat\.

In the particular case of cokriging, *Visual_Data* is rigid with respect to the format of the primary data. An unlimited number of variables, *M*, may be considered, but these variables must occur in order at the beginning of each line of the data file. Moreover, the final two or three values on each line must be in this order: *y*-coordinate, followed by *x*-coordinate (followed by *z*-coordinate for three-dimensional analyses). A suggestion for using *Microsoft's Excel* program to reconfigure data files is presented in Section 6.5.3.

If applying cokriging to a set of data for the first time, variograms of the *M* variables, and cross-variograms for all possible combinations of two variables must first be obtained. This step can be accomplished as soon as the primary data are opened in the main program window. An analyst clicks Tools on the main menu, then chooses cross-variogram analysis. (Note: this tool first computes the variograms of the *M* variables, then computes $M(M - 1)/2$ total cross variograms). This tool prompts users for responses to the following queries:

1. Have you opened a data file? Whereas this seems like a nonsensical question, even experienced analysts can forget to open a data file before choosing a type of analysis.

2. Enter number of variables considered. This value is *M*, the number of different, spatially coexisting variables to be treated in cokriging. In the case of the *Nevada_Landsat* data, *M* is 7.

3. Enter option for directional calculations. Enter 1 for omnidirectional, 2 for a user-specified direction, 4 for a four-directional analysis (EW, NE-SW, NS, and NW-SE), or 8 to obtain a higher resolution analysis of these four directions. If a user responds with 2, user specified direction, *Visual_Data* requires responses to the following queries: A. the direction to be considered; and B. the tolerance, or allowed deviation from this direction to accommodate spatially irregular sampling; for example, a deposit is known to be structurally controlled along a fault having a strike, N30W. This direction is entered into *Visual_Data* as: 120. (EW is 0.; proceeding counterclockwise, NE-SW is 45., NS is 90; consequently N 30 W is equal to 90 + 30, or 120.). Suppose that a tolerance of 20 degrees is used; orientations between N 10 W and N 50 W will consequently be considered.

4. Enter data transform option: 0 for no transform, 1 for log-normal transform, or 2 for indicator transform. If choosing an indicator transform, a user will be asked to specify the threshold (cutoff) for effecting the transform.

5. Define the dimensionality of the analysis: 1, 2, or 3 dimensional data.

6. Define the class size for variogram and cross-variogram calculation; the concept of class size in cokriging is identical to that which is discussed in Chapter 5 for kriging. In the case of the *Nevada_Landsat_1x* data, a class size of 1 or 2 (pixels) is appropriate; for the *Nevada_Landsat_6x* data, 6 (pixels) is the minimum sampling resolution, so a class size of 6, or 12, is more appropriate.

7. Is your data format Simplified GeoEAS? This data format is discussed in Chapter 5, kriging. For more information about it, please see Deutsch and Journel (1992; 1998).

8. Do you wish to rescale your variables, each to a numerical range, [0,1]? This is optional. Earlier in this chapter, rescaling is shown to have no effect on the outcome from cokriging. This option can, however, significantly improve the ability to resolve proper cross-covariance relationships, especially when a large disparity exists in numerical magnitude for a pair of variables. For the *Nevada_Landsat_data*, rescaling is unnecessary in this regard. For the Jura data, however, the outcome from variogram and cross-variogram calculation is more sensitive to rescaling.

For example, the following two figures show the cross-variogram relationship for variables 6 and 7 of the Jura data set. The figure on the left was computed without data rescaling, whereas the figure on the right resulted from data rescaling. These are not identical calculation results. The implication for cokriging is worth noting.

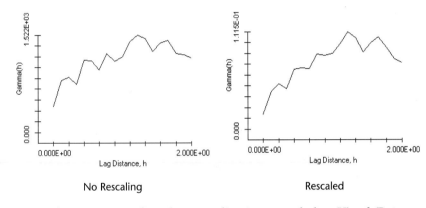

No Rescaling   Rescaled

Once the query regarding data rescaling is responded to, *Visual_Data* completes the calculation of M variograms and $M(M-1)/2$ cross-variograms. A progress bar charts the program's progress through these calculations. Graphs of all variograms and cross-variograms are displayed, one at a time. Moreover, numerical results of each variogram calculation are written to the lower box of the main program window in *Visual_Data*. From File on the main window, an analyst chooses SAVE AS to save these results to a file. Once saved, these results can be accessed later, either by *Visual_Data* or some other program, to create graphs once again of variogram results.

Upon completion of variogram analysis, cokriging is accomplished using *Visual_Data* by first creating two files, one for variogram and cross-variogram information and one for grid information. The same primary data file opened for variogram and cross-variogram calculation is also opened to start cokriging. Once opened, an analyst clicks Tools on the main menu, then moves the mouse cursor to Cokriging. Five suboptions then appear: 1. Create (or edit) grid parameter file; 2. Create (or edit) variogram and cross-variogram parameter file; 3. Proceed to estimation; 4. form a bitmap digital image from gridding results; this tool is used if an analyst chose not to create a bitmap image in the original cokriging, but later has decided to do so by recalling the file of grid results; and 5. form a multiple indicator data set; this fifth tool is used to facilitate indicator cokriging by forming multiple indicator transforms using different cutoffs, but for only a single spatial phenomenon, such as copper, or gold, or benzene contamination.

Examples of variogram/cross-variogram and grid parameter files are shown for the *Nevada_Landsat_6x* data. These files are pasted into this text in the exact form as was used to obtain results shown earlier in this chapter:

```
Variogram/Cross-variogram   parameter   file,   Nevada_
Landsat_6x data:

    3           0.004       0.02       180.    0.    1.
    3           0.004       0.018      180.    0.    1.
    3           0.004       0.02       180.    0.    1.
    1           0.003       0.012      240.    0.    1.
    1           0.004       0.022      180.    0.    1.
    1           0.005       0.037      180.    0.    1.
    1           0.005       0.028      156.    0.    1.
    3           0.015       0.074      180.    0.    1.
    3           0.015       0.077      180.    0.    1.
    3           0.008       0.049      180.    0.    1.
    1           0.015       0.075      200.    0.    1.
    1           0.01        0.093      192.    0.    1.
    1           0.02        0.092      216.    0.    1.
    3           0.015       0.076      220.    0.    1.
    1           0.008       0.049      180.    0.    1.
    1           0.015       0.074      200.    0.    1.
    1           0.009       0.090      192.    0.    1.
    1           0.009       0.090      220.    0.    1.
    1           0.010       0.054      240.    0.    1.
    1           0.016       0.081      200.    0.    1.
    1           0.009       0.095      192.    0.    1.
    1           0.020       0.097      220.    0.    1.
    1           0.015       0.061      200.    0.    1.
    1           0.007       0.070      180.    0.    1.
    1           0.014       0.070      200.    0.    1.
    1           0.0095      0.095      204.    0.    1.
    1           0.020       0.100      216.    0.    1.
    1           0.011       0.110      204.    0.    1.
    1    100000.            5
```

```
Grid   Parameter   File,   Nevada_Landsat_6x   data   (to
obtain grids of 7 variables):

    1
    0
    7
    1        100        1      100     0    0
    1.       -1.        0.
    0.5      0.5        0.
```

When an analyst chooses either option, create variogram/cross-variogram parameter file or create grid parameter file, a window appears that explains how to create these files. This is a similar design to that described in Chapter 5 for kriging.

Once the variogram/cross-variogram and grid parameter files are created, an analyst can proceed to estimation, the third choice on the cokriging
sub-menu. Once this choice is made, the following queries require responses:

1. Have you opened a data file?
2. Is this an undersampled problem, with zeros representing missing
   values? Recall that cokriging can be applied to M variables, in which
   the sampling resolutions of these variables can differ. In fact, this
   may be the only case in which cokriging offers an improvement over
   kriging. If this is an undersampled problem, a zero is entered for
   each missing value, and an analyst responds yes to this query.
3. Select the filename storing the Grid Parameters.
4. Select the filename storing the Variogram and Cross-variogram
   parameters.
5. Is the file format Simplified GeoEAS? Respond yes or no.
6. Enter 1 for cross-validation, or 2 for gridding.
7. Do you wish to rescale variables to a numerical range, [0,1]. This
   choice must be consistent with the variogram and cross-variogram
   parameter file. If these parameters are those obtained using rescaling, then rescaling should be used in cokriging.

Once the seventh query is responded to, cokriging begins.

6.5.2   Using MATLAB?

Frustration was encountered when attempting the use of MATLAB version 5,
for cokriging. Until now in this text, the book by Middleton (2000) has been
the resource for MATLAB applications. In many cases, the author modified
Middleton's programs to fit a particular analysis. Nonetheless, given the author's inexperience with MATLAB, many of these analyses would not have
been completed successfully without the Middleton (2000) resource as a
guide. Middleton (2000), however, does not report a MATLAB program for
cokriging. Whereas Middleton (2000) discusses cokriging in brief, readers
are referred to a program presented by Marcotte (1991) for actual analytical
experiments. This is when the author encountered frustration.

Marcotte (1991) presents a detailed listing of a MATLAB program for cokriging. This particular paper appeared in the journal, *Computers and Geo-
ciences*, a publication of the International Association for Mathematical
Geology (IAMG). Many of the programs reported in C&G are available for
downloading from the IAMG web site: http://www.iamg.org. Unfortunate-
ly, the MATLAB program by Marcotte (1991) is not available for download.

Consequently, a user is confronted by the necessity of entering this program, line by line, using the listing in the journal article. This is a nontrivial
task given the length of this program. Moreover, because Marcotte (1991) uses
lower case entry for many variable names, the variable l (lower case L) looks
identical to 1 (the number, one). Furthermore, MATLAB has progressed through
several versions since this article appeared in 1991. Some syntax changes have
occurred. For instance, Marcotte (1991) uses a MATLAB command, casesen off.
This command caused a syntax error in MATLAB *Version 5.3*. More significantly,
the following command used by Marcotte (1991) caused a syntax error:

```
(in cokri, Marcotte (1991), line 120):
grid = t2.*(ones(ng,1)*block)
```

MATLAB (*Version 5.3*) considered these [matrices] to be of incorrect size for the proper execution of this command. This is not the first time the author encountered this error when using MATLAB. In previous chapters, when modifying Middleton (2000) code, similar errors occurred. MATLAB is supposed to offer a shorthand way to multiply two matrices. For instance, if and *b* are matrices, then

```
>> c = a * b
```

should yield a matrix, *c*, that is the product of *a* and *b*. Matrices, *a* and *b*, must be declared as matrices, moreover their sizes must allow the product, *c*, to be properly obtained. In many cases, the author has encountered an error with this syntax, with MATLAB returning the error message indicating the matrices are of an incorrect size. The method around this error used frequently by the author involves explicitly coding the matrix algebra:

```
for i = 1:n,
        for j = 1:m,
                c(i,j) = 0.
                for k = 1:p,
                        c(i,j) = a(i,k) * b(k, j)
                end
        end
end
```

MATLAB is supposed to alleviate the need for this type of explicit coding, and it is the inexperience of the author that most probably explains the frustration. With Middleton (2000), given the short nature of the codes, the author could easily design an explicit coding solution to each syntax error. With Marcotte (1991), however, given the length of the code, the author was not able to design an explicit solution to syntax errors.

Perhaps MATLAB is more useful for simpler applications. Marcotte (1991) is a complicated, lengthy code. MATLAB is also supposed to minimize coding, yet Marcotte (1991) is almost as complex as a *Fortran* or *Visual Basic* algorithm for cokriging.

6.5.3   Using *Microsoft Excel?*

Excel applications are excluded from the chapter on kriging, Chapter 5, because using this software for kriging is complicated and involved, or is not possible. The same statement applies to cokriging. The intent of this section, however, is to show how *Excel* can be used to enhance the process of cokriging using another piece of software, such as *Visual_Data*.

Earlier in this chapter, a demonstration is presented showing how Landsat TM band 7 (mid-infrared) data can be used to enhance the resolution of Landsat TM band 6 (thermal) data. This is a two variable ($M = 2$) case. The original *Nevada_Landsat_data* consist of seven variables. A data set containing only the sixth and seventh variables was constructed using *Microsoft Excel*. The *Nevada_Landsat_1x* data (full resolution data) were imported into *Microsoft Excel*. The first five columns corresponding to the first five variables were deleted, leaving four columns: column A = variable 6 (thermal), column B = variable 7 (near infrared), column C = y-coordinate, and column D = x-coordinate. This amended file was then saved under a new name.

Likewise, *Excel* is useful for rearranging a data file to match the input requirements of *Visual_Data*. Recall that *Visual_Data* expects the primary data file for cokriging to be arranged in this order:

```
Variable 1, Variable 2, ... Variable M, Y-coordinate,
X-coordinate, [optional: Z-coordinate].
```

If not in this order, *Excel* can be used to rearrange columns.

6.6   Literature

This chapter is about **generalized cokriging,** a technique fully developed and discussed by Myers (1982). The method of paired-sums for cross-variogram computation is that which is used by Myers (1982) for generalized cokriging. This chapter has involved many experiments to compare cokriging to kriging. In most of these, cokriging did not yield an improvement over kriging. Wackernagel (1995) is the inspirational work for explaining this outcome. The notion of autokrigeability discussed earlier in this chapter is borrowed directly from Wackernagel (1995). The Jura data used in this chapter are taken as listed in Goovaerts (1997). Finally, Marcotte (1991) is the fundamental resource for a MATLAB environment for cokriging. It is this author's inexperience with MATLAB that prevented a successful application.

6.7   Final Thoughts on Cokriging

Autokrigeability is a fundamental aspect of this chapter. If data are autokrigeable, and this is the case if data have an intrinsic autocovariance, then cokriging yields results that are identical to kriging. This chapter also argues this case from a different perspective. If the sampling resolutions for the M variables are identical, then no improvement is possible with cokriging. As is explained in Chapter 8 on digital image processing, applying weights that sum to zero to spatial data in a local neighborhood yields an interpolation (estimation) that is a high-pass filter of the original data. In other words, the estimated value represents the high frequency (high spatial frequency) information content of the original data. In cokriging, estimation proceeds as follows (a qualitative review):

```
Estimate, variable i = W * i + P * j
```

In which i is the primary variable and j is the metaphor for secondary information. The weights, W, add to 1, whereas the weights, P, add to zero. If i and j are sampled at the same resolution, the high frequency content of i matches that of j and no benefit is obtained from j when estimating i. This is another way to understand autokrigeability.

If, on the other hand, the resolution of j is better (higher) than that of i, then j has more high frequency information than does i and an improvement is possible using cokriging. This is the undersampled case in cokriging. A demonstration is presented in this chapter in which the resolution of the thermal image in a collection of Landsat TM images is improved using one of the higher resolution images. An improvement from cokriging is only possible in the case of undersampling. Otherwise, if the sampling resolutions of all M variables are equal, kriging the M variables separately yields the same accuracy without the need for modeling cross-variograms.

Exercises

1. Review section 6.7 again—Final thoughts on cokriging. Consider the following sampling geometry. Show for any variogram and cross-variogram models that cokriging and kriging yield the same result (estimate) for variable 1 at location, 0.

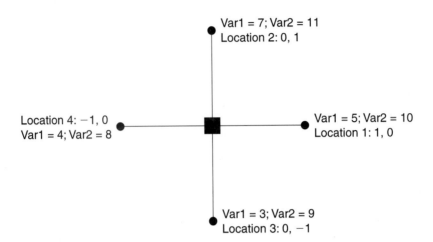

Var1 = 7; Var2 = 11
Location 2: 0, 1

Location 4: −1, 0
Var1 = 4; Var2 = 8

Var1 = 5; Var2 = 10
Location 1: 1, 0

Var1 = 3; Var2 = 9
Location 3: 0, −1

2. Again using the sampling geometry shown in problem 1, assume variable 1 is not sampled at location 2 (change its value to 0). Show that cokriging and kriging now yield different estimates at location 0.

3. Instead of using the paired-sum approach to obtaining a cross-variogram, many texts present a direct method for computing the cross-variogram (e.g., Chiles and Delfiner, 1999, p. 321):

$$\gamma_{ij}(h) = \frac{1}{2N} \sum_{k=1}^{N} [Z_i(x) - Z_j(x + h)]^2$$

Using the following data, solve this equation by hand. Then, use the paired-sum approach, solving by hand for the paired-sum variogram. Then, solve for the cross-variogram as:

$$\gamma_{ij}(h) = 0.5[\gamma_{ij}^+(h) - \gamma_i(h) - \gamma_j(h)]$$

rewritten qualitatively as: *cross-variogram* = 0.5(paired sum variogram − variogram, i − variogram, j). By hand, in other words, three variograms are computed: paired sum, that for variable, i, alone, and that for variable, j, alone.

How do the two results compare? What is the significance of similarities or differences? What is the relationship between the two approaches?

Data for problem 3:

I	J	Y	X
181	90	1	1
182	91	1	2

I	J	Y	X
182	93	1	3
181	93	1	4
182	95	1	5
182	94	1	6
182	93	1	7
181	91	1	8
181	88	1	9
181	87	1	10
181	88	1	11
180	87	1	12
179	87	1	13
180	82	1	14
179	82	1	15
179	85	1	16
178	83	1	17
178	81	1	18
179	85	1	19
179	88	1	20
179	88	1	21
179	89	1	22
178	89	1	23
178	90	1	24
177	90	1	25
177	91	1	26
177	90	1	27
177	90	1	28
178	91	1	29
178	91	1	30
179	92	1	31
179	92	1	32
180	91	1	33
181	88	1	34
181	86	1	35

Spatial Simulation

7

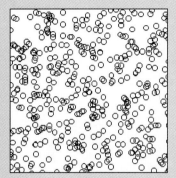

In the previous two chapters, the focus is on *estimation*. Now, the focus shifts to *simulation*. What is the difference? Kriging is applied to obtain the best possible estimate of the value of a spatial phenomenon at an unsampled location. This estimate is unbiased, and is based on information defining spatial autocovariance. Kriging, nonetheless, is a low-pass filter; that is, it smooths data. The histogram and variogram of all estimates, Z*, differ from those for the original data. In contrast, simulation is not the best possible estimation method. Its purpose is the modeling of a spatial phenomenon to yield an outcome that has the same histogram and variogram as the original data. Simulation is the best method for modeling the *variability in space* of natural phenomena.

Simulation implies artificiality. It is a game, a fantasy. Yet, here is a small sampling of its virtues. The sulfur values in a coal seam are modeled using simulation. Information used to develop the simulation is obtained from actual sulfur data. Once the simulated model is developed, it is "mined" as if it is an actual coal seam. The goal is to better appreciate the fluctuation in the amount of sulfur an electrical utility company can anticipate when burning coal from this particular seam. Engineers are then able to design a better emissions control system (scrubber) to minimize sulfur emissions. Another example is taken from the precious metal mining industry. A deposit of gold is simulated having the same histogram and variogram as the actual deposit. This simulated model is then "mined" to appreciate the fluctuation in gold grade that can be expected on a daily basis. The purpose is to understand how much variation in efficiency that can be expected from mill production for this particular deposit. Finally, a university wishes to sell its land that had been used in the past for training fighters of range and forest fires. The site, though, is contaminated by benzene and must be cleaned before close of sale. Field sampling has identified some benzene "hot spots," but how many more hot spots may be encountered that are unsampled? Benzene amounts are simulated at this site and randomly sampled to compare to field sampling. The goal is to better appreciate how much more benzene may be present that is not sampled.

These are but a few of many possible applications of simulation. In this chapter, we will use simulation to model the variability in pixel values exhibited by the *Nevada_Landsat* data. Simulation is based on random number generation. This chapter consequently opens with random number visualization.

7.1 Random Numbers and Their Generation

This discussion is developed from Pickover (1995).

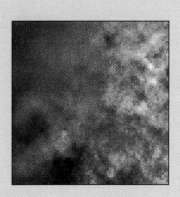

Consider the number known as π:

3.14159265358979323846264338327950288419716939937510582097494 ...

and further consider only that portion of π *after* the decimal:

.141592653589793238462643383279502884197169399375105820974944 . . .

This sequence of numbers is random! In other words, consider each digit after the decimal as a random number. One necessary characteristic of random digits is that they occur with equal frequency. Consider the roll of a die with six possible outcomes. If the die is fair, any one of these six digits has an equal chance of occurring. A fair die is consequently a **random number generator.** Can π be used as a random number generator? A histogram of its digits after the decimal suggests that it can.

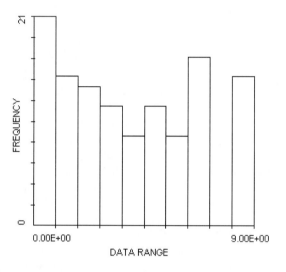

Treating the fractional portion of a real number after the decimal as a random number stimulates the imagination of a large number of possible generators. For example, consider the following generator (Pickover, 1995, pp. 235–236, *Cliff RNG*):

```
x = 0.1
for i = 1 to num
    x = 100 * log(x)
    ix = Int(x)
    x = Abs(x-ix)
next i
```

The absolute difference, $Abs(x - ix)$ yields that portion of the functional outcome after the decimal because x is the real number and ix is only the whole number portion of x. A *Visual_Basic* program is available on the CD-ROM called RNG_Fun. Experimenting with the Cliff RNG is one of its options, with the resultant visualization of the randomness of its numbers.

This visualization is developed by generating 1000 pairs of random numbers, one defining the x-coordinate direction and the other defining y, and drawing a circle at the point, (x,y). Figure 7.1 suggests that the quality of the Cliff random number generator is very high with respect to randomness.

Notice what happens, though, if we change the Cliff RNG slightly as follows:

rather than using $x = 100 * \log(x)$,

we instead use $x = \log(x)$.

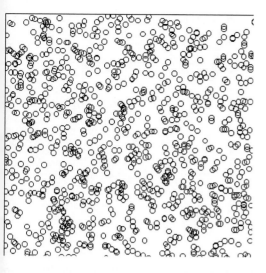

◀ FIGURE 7.1
Outcome from the Cliff random
number generator.

This yields Figure 7.2. Notice the effect of the multiplier. This shows that
the fractional portion of a functional outcome is a random digit only in certain circumstances.

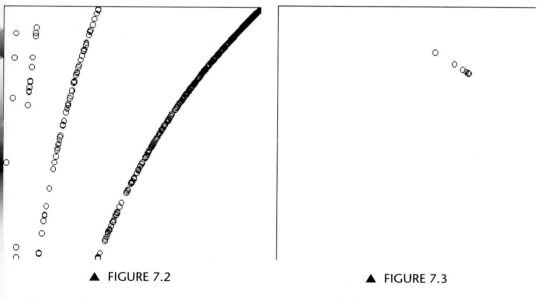

▲ FIGURE 7.2   ▲ FIGURE 7.3

Notice that the numbers generated using the modified Cliff RNG are distinctly nonrandom. This modification is a poor random number generator.

Pickover (1995) suggests that it is a rather easy matter to choose any
function and use it as a random number generator. This challenge prompts
the following experiment:

$x = Cos(x)$ as a random number generator, starting with the initial value,

$$x = 0.4.$$

The visualization in Figure 7.3 shows the consequence.

This is a poor random number generator. The function is modified accordingly to become:

$$x = 100 * Cos(x), \text{ starting with } x = 0.4.$$

The following visualization shows the consequence (Figure 7.4):

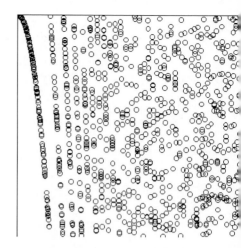

▶ FIGURE 7.4
Notice the nonrandom behavior on the left-hand side.

This is still a poor generator, but the visualization shows more random ness in comparison to the first experiment, suggesting that the modification was in the right direction. A further experiment was attempted:

$$x = 1000 * \text{Cos}(x), \text{ starting with } x = 0.4;$$

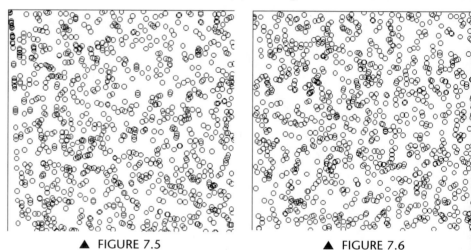

▲ FIGURE 7.5   ▲ FIGURE 7.6

and this is a better outcome (Figure 7.5). But, careful examination of the upper, left corner of the visualization indicates that there is still some nonran domness in the outcome. A final experiment yields a high quality random number generator (Figure 7.6):

$$x = 10000 * \text{Cos}(x), \text{ starting with } x = 0.4:$$

Pickover's challenge is successfully met. The matter of designing a ran dom number generator is actually quite simple.

7.2   One-Dimensional Spatial Simulation

Random number generation is the essential first step in spatial simulation. Con sider one-dimensional space. The next step requires processing random num bers to become **spatially correlated random digits.** This step is not daunting and is best demonstrated using what is known as a **moving window average.** Recall the portion of π after the decimal:

.141592653589793238462643383279502884197169399375105820974944 . . .

and suppose we pass a 1 × 3 moving window along this string of numbers, av raging the values each time that are within the window, incrementing the window by one digit along this string of numbers. This is demonstrated for clarity:

1. start with the numbers, 1, 4, and 1, and average them: $1 + 4 + 1 = 6/3 = 2$
2. move the window 1 digit to the right and average these three numbers: 4, 1, and 5: $4 + 1 + 5 = 10/3 = 3.33$. Notice that two of the numbers used in step 1, 4 and 1, are also used in step 2.
3. move the window again 1 digit to the right, and average the three numbers:
   1, 5, and 9: $1 + 5 + 9 = 15/3 = 5$; notice that one of the numbers, 1, has been used in steps 1, 2, and 3;
4. move the window again and average:

$$5 + 9 + 2 = 16/3 = 5.33$$

Step 4 is repeated until there are no numbers remaining. A variogram omputed using the processed numbers is compared to the variogram for he original, unprocessed random digits to show the outcome:

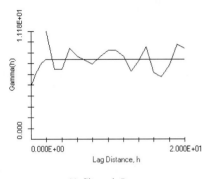

Unfiltered, Raw

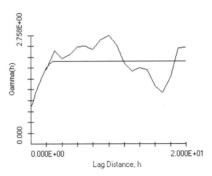

Filtered, 3 × 3 window

Notice that the moving window average yields spatially correlated numbers aaving a range equal to the size of the moving window, 3 in this case. Obaining a simulation having a variogram with range, r, simply involves the election of a moving window, size $1 \times r$.

Further processing of the averages is necessary to obtain the desired variance (sill) and noise (nugget). Random numbers are generated in the range, 0,1]. In spatial simulation, we actually desire a first simulation that is modeled vell by a normal distribution and has a mean of zero and variance of one (zero nean, unit variance is a commonly referred to concept in geostatistics). Obtaining a simulation having a zero mean is straightforward. Once a random digit is obtained from the random number generator, simply subtracting 0.5 from it gives a random number in the range, $[-0.5, 0.5]$, having a mean of zero. Obaining the requisite variance requires the imposition of the appropriate variogram model (spatial law), spherical, exponential, and so on. To obtain a unit variance with a spherical spatial law, the numbers are processed as follows:

$$NR = \text{Range}/(2 * \text{Grid Spacing});$$

$$\text{then } w = SQRT(36/(NR*(NR + 1)*(2NR + 1)))$$

hen:

$$y_i(J) = w \sum_{m=-NR}^{NR} mT(J + m)$$

in which y are the processed random digits having a spherical covariance, zero mean, and unit variance and T are the random digits generated by the random number generator. Notice that these steps incorporate the moving window averaging, with window size, $2NR + 1$.

7.3 Extension to Three-Dimensional Space: The Method of Random Lines

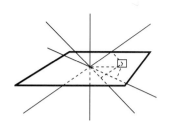

▲ FIGURE 7.7
Three-dimensional simulation effected as N one-dimensional simulations projected onto the equatorial plane.

A number of one-dimensional simulations are used to develop a simulation within a three-dimensional volume. Most typically, a two-dimensional simulation is desired. In this case, a sphere is imagined (three-dimensional space) and a two-dimensional simulation is developed as a projection from each one-dimensional simulation onto the equator (Figure 7.7).

Each one-dimensional simulation is treated as a vector oriented in three-dimensional space. Each of these vectors passes through the center of the sphere (Figure 7.7). The orientation is represented as $\{x', y', z'\}$, where x', y', and z' are obtained as random digits in the interval, $[-1, 1]$, using a random number generator, subtracting 0.5 from the outcome, then multiplying this outcome by 2. This orientation algorithm is the nucleus of the **method of random lines**.

In this method, N one-dimensional simulations are used, where N can be any number. The algorithm is as follows:

Step 1: define the equatorial plane of the sphere as an $NROW \times NCOL$ grid (Figure 7.7).

Step 2: create N one-dimensional simulations, each of length $[NROW^2 + NCOL^2]^{1/2}$;

Step 3: orient the N vectors by generating x', y', and z' for each;

Step 4: for each grid cell, (i,j), find the closest value, y, on each of the N vectors, then compute the simulated value as a global sum: $[\sum y]/[N]^{1/2}$.

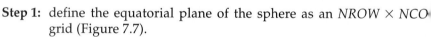

Although the process appears rather simple, simulation in practice requires experimentation to obtain a model of the most desirable quality. In the random line algorithm, whereas N can be any number, the larger N is, the better is the visual outcome of the simulation. Small N yields simulations that are noticeably striped, consequently appearing nonrandom. Several visualizations are presented demonstrating the influence of N on the simulated outcome (Figure 7.8).

Striping is severe for $N = 10$ lines. Moreover, striping is still evident when $N = 50$ lines. The outcome when $N = 200$ lines does appear random and striping is not evident. Some striping seems evident in the outcome for $N = 1000$ lines. If this is a real artifact, and not an illusion, then there may be a limit to N, beyond which striping again becomes evident.

Rather than offering speculation, this notion is tested explicitly by using $N = 2000$ and $N = 3000$ lines (shown on the bottom of the next page). From a visual perspective, these two results are similar to that for $N = 1000$ lines. There are some linear artifacts that are probably related to the algorithm. Moreover, given the visual similarity of the $N = 1000, 2000$, and 3000 results, there does not appear to be an advantage to using more than 1000 lines in the random lines algorithm. Chiles and Delfiner (1999) remark that N should be at least "one to several hundred." lines (p. 477), implying an upper limit to N. This limit, if the implication is indeed understood correctly, is consistent with the experiments presented in this section.

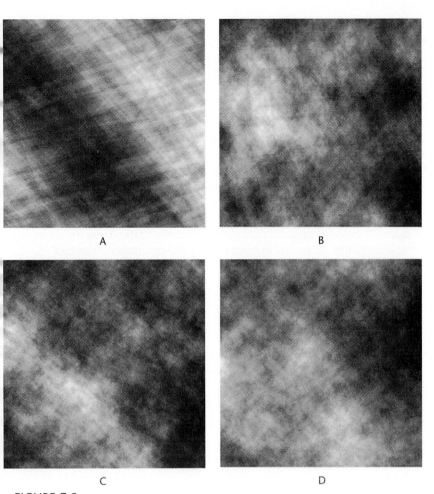

FIGURE 7.8

Four experiments to show the visual influence of N, the number of random lines, on the simulated outcome: A. 10 lines; B. 50 lines; C. 200 lines; and D. 1000 lines. Notice that when using fewer number of lines, N, a noticeable striping artifact contaminates the simulation.

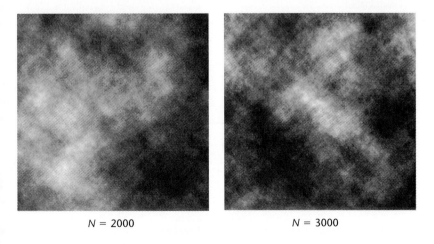

7.4   Simulation Using Fractals

An alternative simulation algorithm is based on the notion of fractals. This notion is introduced in Chapter 5. Therein, fractal dimension was shown to be an equivalent concept to the variogram. In fact, fractal dimension, D, is equal to $2 - (\log(\gamma)/2\log(h))$. Fractal dimension, D, is also equal to Hurst dimension, H, as $2-H$.

Carr (1995) presents *Fortran* programs for one and two-dimensional simulation based on fractals. These computer algorithms, developed using Peitgen and Saupe (1988), are based on H, the Hurst dimension. The closer H is to one, the smoother, more regular the simulated outcome. Smooth topography, for instance, has a Hurst dimension in the range, [0.8 to 0.9]. The closer H is to zero, the more random, or rougher, the simulated outcome. Fracture surfaces in metals, for instance, may have Hurst dimensions in the range [0.1 to 0.2].

Random number generation is used to effect a simulation in the **frequency domain.** Two-dimensional, inverse Fourier transformation (recall Chapter 5) is used to transform the frequency domain simulation to the spatial domain. Several visualizations are presented to illustrate the influence of Hurst dimension, H, on the appearance of the simulation (Figure 7.9).

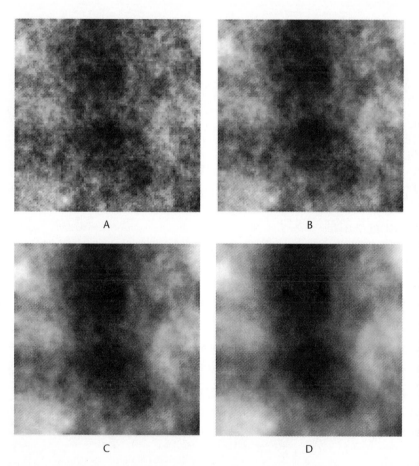

▲ FIGURE 7.9
Four simulations based on fractal dimension (Hurst dimension). Simulations are based, in order, on: A. $H = 0.2$; B. $H = 0.4$; C. $H = 0.6$; D. $H = 0.8$. The larger H is, the smoother is the outcome. All four simulations were obtained by starting the random number generator (Cosine generator) at the same point, the 2001 outcome. Unlike the random line algorithm, in which the total number of calls to the random number generator is a function of the number of lines used to effect the simulation, the fractal-based simulation uses the same number of calls to the random number generator, regardless of the value, H. Thus, if the random number sequence is started at the same point for each simulation, the pattern of numbers will be the same, consequently the simulations will have a similar appearance. Notice that these four simulations look similar, except the smoothness changes with H.

.5   Nonconditional Simulation: The Need For Data Transformation

oregoing simulations are examples of **nonconditional simulations.** Such mulations share the same variogram with sample data, but do not honor ιe data at their sampling locations except by coincidence. The simulations resented in Sections 7.3 and 7.4 are constrained by their algorithms to fair •presentation by the normal distribution model, and have mean values of ero and variances equal to one (a so-called **mean zero-variance one process**). ι this sense, nonconditional simulations do not share the same histogram ith sample data, except by coincidence.

A transform is necessary for a nonconditional simulation's histogram to ιatch that for sample data. Two transforms are offered in *Visual_Data*, a sim-ιe, linear transform and a more powerful, but complicated one that is based ι Hermite orthogonal polynomials. The simple, linear transform is imple-ιented as

$$Z_s^T = Mean + Z_s\sqrt{Var}$$

[the transformed simulated value equals the desired mean value plus the original, nonconditional simulated value multiplied by the square root of the desired variance]

. linear transform achieves the desired mean and variance. But, it preserves ιe normality of the original simulation. This is fine if the original spatial ιhenomenon that is being simulated is also represented well by a normal ιstribution model. But, this type of transform is not applicable to data for ιhich the assumption of normality is poor.

A more versatile transform is afforded using Hermite orthogonal poly-ιmials. This transform can yield any distribution model, consequently it is ιpable of modeling many different data distributions. This transform is a ιcursive algorithm as follows:

$$C_j = Coef_j + Z_sC_{j+1} - jC_{j+2}, j = Nterm, Nterm - 1, \ldots, 1 \text{ by } -1$$

ι which *Coef* are coefficients of the Hermite polynomials and *Nterm* is the ιumber of terms desired in this polynomial representation. Once the values, , are computed, then

$$Z_s^T = C_1$$

ι practice, a sufficient transform is effected using *Nterm* = 5; more terms ιan this can lead to numerical instability, a conclusion that is based on ικperience.

A reference is made in Chapter 2 to the Hermite polynomial transform. ιherein, the coefficients, *Coef*, are discussed in the context of statistical mo-ιents. Indeed, $Coef_1$ is equal to the desired mean value of the transformed ιmulation. The other coefficients are proportional to higher statistical mo-ιents. For example, the second Hermite coefficient is proportional to the ιquare root of statistical variance (standard deviation).

DEMONSTRATION

Hermite polynomial coefficients for the first five terms are computed for the visi-ble blue reflectance, *Nevada_Landsat_6x* data. These coefficients are computed using the Histogram tool in *Visual_Data*. Coefficients are written to an output file for use in simulation experiments. Results are as follows:

Term	Coefficient
1	148.4538
2	23.3941
3	−2.9305
4	−0.1753
5	0.6053

These coefficients are used in simulation experiments that are presented in the next several sections. Calculation Example. A nonconditional simulation prior to transformation has a mean of zero and variance of 1. Suppose this simulation is transformed using a Hermite polynomial expansion based on the coefficients for the visible blue reflectance, *Nevada_Landsat_6x* data. What is the transformed equivalent of a simulated value equal to 0.5?

Following the Hermite polynomial expansion algorithm shown earlier, the following steps are executed:

$$C_5 = Coef_5 + Z_sC_6 + jC_7 = 0.6053 + 0.5 * 0.0 + 5 * 0.0 = 0.6053$$

$$C_4 = -0.1753 + 0.5 * 0.6053 + 4 * 0.0 = 0.12735$$

$$C_3 = -2.9305 + 0.5 * 12735 + 3 * 0.6053 = -1.051$$

$$C_2 = 23.3941 + 0.5 * (-1.051) + 2 * (0.12735) = 23.12$$

$$C_1 = 148.4538 + 0.5 * (23.12) + 1 * (-1.051) = 158.96$$

The transformed equivalent of 0.5 is equal to C_1, 158.96.

Transforming a nonconditional simulation is essential when modeling data distributions. Algorithms for nonconditional simulations produce numerical values that conform precisely to a normal distribution model. Many natural phenomena are associated with data that are not represented well by a normal distribution model. Transformation is necessary for a nonconditional simulation to be a representative model of any one of these natural phenomena.

7.6   Nonconditional Simulation of *Nevada_Landsat_Data*

Several experiments are presented to demonstrate the nonconditional simulation of natural phenomena. Two such phenomena are chosen, visible blue reflectance and thermal emission, *Nevada_Landsat_data*. Because we learned in Chapter 2 that a normal distribution model does not fit the distributions of these data well, Hermitian transformation is compared to linear transformation to show the importance of modeling actual data distributions as closely as possible when developing models. Both transforms are also applied to compare and contrast the influence of transform algorithm on resultant visualizations.

7.6.1   Random Lines Algorithm

In these experiments, the number of lines used to develop simulations is arbitrarily set to 500. We learned earlier in this Chapter that using 1000 or more lines offers little, if any improvement to the visual appearance of the simulation, but using too few lines yields simulations with noticeable, linear artifacts. The choice of 500 lines is a compromise. Square simulations are effected of size, 100 × 100, for comparison to 100 × 100 actual subimages from the Landsat images. Nonconditional simulations will never look like actual data. Comparison of simulations to actual data, though, helps ana

sts to visualize natural spatial variability to judge the quality of simulation
odels for representing the pure essence of spatial autocovariance.

6.1.1   Visible Blue Reflectance

ie variogram for visible blue reflectance (Chapter 5) showed a range of at
ast 240 pixels. A simulation model is developed to honor this range. This
mulation is shown in the following steps: 1. nonconditional simulation, no
ansform; 2. nonconditional simulation, linear transform, mean = 148.45,
iriance = 592.8; and 3. nonconditional simulation, Hermitian transform
sing the coefficients shown in Section 7.5:

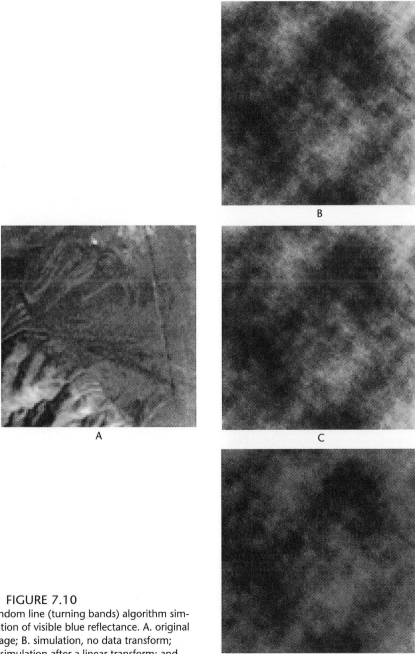

FIGURE 7.10
ndom line (turning bands) algorithm sim-
ation of visible blue reflectance. A. original
iage; B. simulation, no data transform;
simulation after a linear transform; and
simulation after Hermitian transformation.

Parameters used to obtain these simulations are: size, N, of the simulation $=$ 100; range of the spatial law $= 240$; number of lines $= 500$; starting point of the random number generator $= 1$.

Notice that the simulation without a data transform is identical in appearance to that obtained using a linear transform. This is an important realization. A linear transform will not alter the distribution of simulated values yielded by the algorithm. Only the values (numbers) are rescaled. In contrast, a Hermitian transform does alter the distribution resulting from the simulation algorithm. The visual appearance of the simulation is consequently altered.

7.6.1.2   Thermal Emission

The spatial covariance properties of the thermal band of the *Nevada_Landsat data* are discussed in Chapter 6. These data are associated with a smaller range in comparison to visible blue reflectance. Modeling this smaller range in simulation experiments yields visually different spatial patterns, even though the same parameters are used to effect simulation (the sole exception is the range of the spatial law, set to 144 in these experiments):

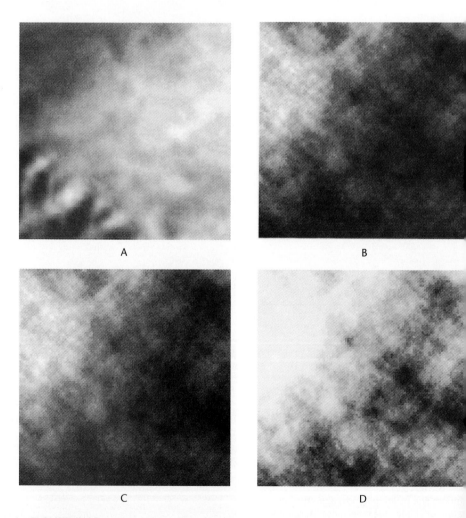

▲ FIGURE 7.11
Simulation experiments applied to the spatial modeling of thermal emissivity. A. actual image; B. simulation, no transform; C. simulation after linear transform; D. simulation after Hermitian transform.

As is the case with the simulation of visible blue reflectance, the simulations of thermal emissivity with no transform and linear transform are identical. Only the simulation after Hermitian transform has a histogram similar to the actual data. This is illustrated by the following three figures:

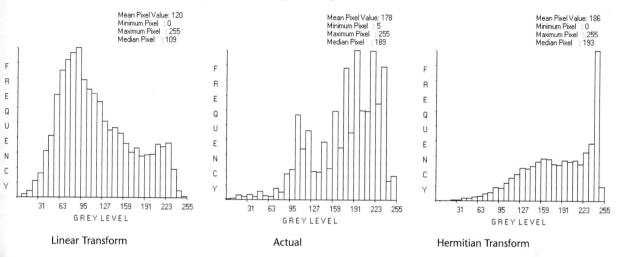

Linear Transform   Actual   Hermitian Transform

Notice that the linear transform is associated with a distribution that is more symmetrical. Both the histogram for the actual data and that after Hermitian transformation are skewed to the left, with the majority of data values tending toward the high data end. In contrast, the linear transform has yielded a majority of values more toward the lower data end.

Variogram reproduction is also an important aspect of geostatistical simulation, and data transform has an influence on this as well. Four variograms are compared to underscore this claim. The actual variogram is compared to that from nonconditional simulation, after linear and Hermitian transform, and conditional simulation after Hermitian transform:

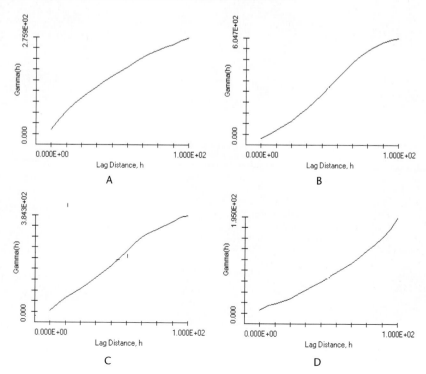

◀ FIGURE 7.12
Four variograms to illustrate the effectiveness of geostatistical simulation with data transform for modeling spatial autocorrelation. A. variogram for the actual thermal data; B. variogram after the application of a linear transform; notice the regular shape of this variogram; C. variogram after the application of the Hermitian transform; and D. variogram of the conditional simulation based on the Hermitian transform.

7.6.2 The Fractal Algorithm

Experiments with the simulation algorithm based on fractal dimension (Hurst dimension) are presented for comparison to outcomes from the random line (turning bands) algorithm. This algorithm requires information defining the Hurst dimension when effecting simulations. This information is obtainable from the variogram. That for visible blue reflectance is shown in Chapter 5. Using the portion of this variogram closest to the origin (the region over smaller h), Hurst dimension, H, for visible blue reflectance is computed as:

$$H = \frac{\log(\delta\gamma)}{2\log(\delta h)} = \frac{\log(107)}{2\log(24)} = 0.735$$

And, for thermal emissivity:

$$H = \frac{\log(40.23)}{2\log(12)} = 0.744$$

The Hurst dimension for thermal emissivity is a little greater than that for visible blue reflectance, indicating a smaller fractal dimension. This is consistent with the lower resolution (smoothed) characteristic of thermal emissivity (in the case of the Landsat TM satellite system).

Based on these values of Hurst dimension, and the following parameters, size, N, of the simulation is 100 and starting point of the random number generator is 1, the following visualizations are obtained:

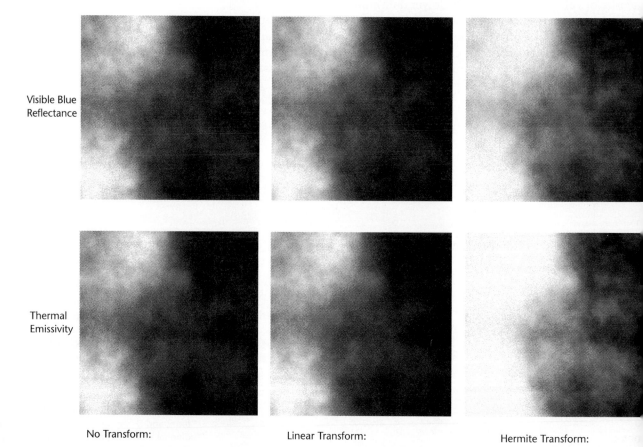

Visible Blue
Reflectance

Thermal
Emissivity

No Transform: Linear Transform: Hermite Transform:

Notice the outcomes for both visible blue reflectance and thermal emisvity. They exhibit identical spatial patterns. Is this a realistic outcome? Then using the fractal-based simulation algorithm, if the random number generator is started in the same place, the same spatial pattern will result, regardless of the value, H, used to generate the simulation. In these experiments, the random number generator is started at 1 for both phenomena. Consequently, the spatial patterns are the same. What differs is the transform, especially the Hermite transform, applied to each outcome. Notice that the Hermite transformed results are more differentiable for these two phenomena in comparison to the linear transformed results.

.7   Conditioning the Simulation

Conditional simulation yields a model that not only shares the variogram and histogram] with a set of spatial data, but also honors these data values at their sampling locations. This is tantamount to stating that the simulation is *conditioned* by (or to) the spatial data. A conditional simulation has an appearance that is more similar to a kriged map, of course because the same spatial data influence the outcome.

Conditioning a nonconditional simulation is a fairly trivial step. Let the actual spatial data be called Z, and further let the nonconditional simulation be called S. The upper, left corner of the simulation (Ymax, Xmin) is defined using the same coordinates as are used to define the spatial locations of Z. Then:

Step 1: at each location, $Z(x)$, the nearest nonconditionally simulated value, $S(x)$ is found; this yields, in total, two data sets: $Z(x)$ and $S(x)$;

Step 2: based on the same variogram for each, kriging is used to condition the simulation as follows; letting C represent the conditional simulation, then

$$C(i,j) = S(i,j) - S^*(i,j) + Z^*(i,j)$$

in which $C(i,j)$ is the conditionally simulated value at grid node, row = i, column = j; $S(i,j)$ is the nonconditionally simulated value at grid node, i,j, $S^*(i,j)$ is the kriged estimate at grid node, i,j, that is obtained using $S(x)$, and $Z^*(i,j)$ is the kriged estimate at grid node, i,j, that is obtained using $Z(x)$.

Kriging is an **exact interpolator.** That is, when kriging is used to obtain an estimate, Z^*, at a location for which a known value, Z, exists, then $Z^* = Z$. When used to condition a simulation, at locations, x, for which Z is known $Z(x)$), then $S(x) = S(i,j)$. In this case, $S^*(i,j) = S(i,j)$, moreover, $Z^*(i,j) = Z(x)$, consequently, $C(i,j) = Z(x)$. Thus, the conditional simulation exactly matches a data value, Z, at a location, x, when grid node, i,j and x coincide. Otherwise, the conditioning step alters all original nonconditionally simulated values such that the variogram of C matches that for Z (and S).

.7.1   Application to the *Nevada_Landsat_Data*

Section 7.6 presents nonconditional simulation experiments applied to the *Nevada_Landsat_data*. These simulations are similar to the actual images in terms of spatial variability, but do not look the same as these images because no conditioning has been applied. In this section, conditioning is used as a final step in the development of simulated Landsat images.

7.7.1.1 Visible Blue Reflectance

In this experiment, only the nonconditional simulation obtained using th random lines algorithm with Hermitian transform is used. Hermitian tran formation yields the closest histogram to the actual data. The random lin simulation is chosen arbitrarily. The outcome is compared to the actu image to show that conditioning renders a visual result similar to the actu data, but with some differences that are attributable to simulation algorith founded on random number generation:

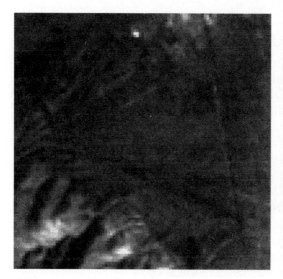
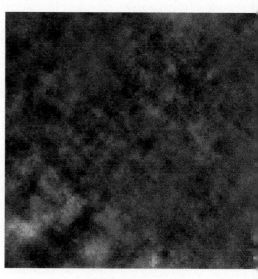

Actual image   Simulated image

The *Nevada_Landsat_6x* data set is used for conditioning, whereas th actual image used for comparison is full resolution. The intent is to show that the simulated model does incorporate details of the actual spatial phe nomenon. It also incorporates details not present in the actual data. group of higher pixel values (light grey to white colors) is noted in th lower, right corner of the simulated image. The simulation indicates tha such a zone of higher pixel values would be consistent with the variograr used to develop the simulation. This is the value of simulation, a game t explore the probability of higher and lower values that are not sampled i the original data set.

7.7.1.2 Thermal Emissivity

A similar simulation experiment is conducted using the thermal emissivit (band 6) information in the *Nevada_Landsat_6x* data set. On the next pag conditional simulation is compared to the actual image:

This is an interesting outcome. The simulation is smoother in th upper, left corner, and more random toward the lower right corner. Th actual thermal data, though, are more regular in the upper, left, and le† edges of the actual image, and are more random toward the lower righ Conditioning, once again, imparts many attributes observable in the actu al data, but also incorporates features not necessarily observable in the ac tual data, although still consistent with the variogram used to develop th simulation.

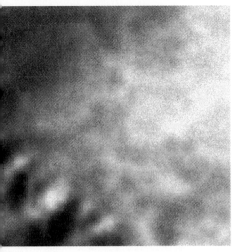

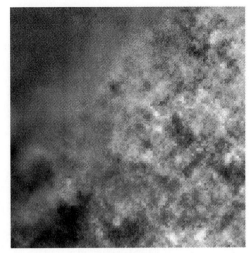

Actual   Simulated

.8   Why Is Spatial Simulation Useful?

imulation is a game. Random number generation is the origin of its numbers.
'onsequently, it is not a method for obtaining the best possible estimate of a
uantity in space. That is the purpose of kriging. Instead, simulation is applied
) experiment with the many possible configurations a spatial phenomenon
ıay exhibit, given the fact that we often can sample it at only a relatively few
)cations. This is the essence of the game, changing the locations of data highs
nd lows while preserving the variogram. These locations are changed by
tarting random number generation at different locations in the sequence. The
)llowing nonconditional simulations demonstrate this influence:

Visible blue, RNG started at 1   Visible blue, RNG started at 10001

Conditioning constrains the game by necessitating conformity to the val-
es of the true spatial phenomenon where it was sampled. Still, because ran-
om number generation is the foundation of the game, locations of highs
nd lows can change, while honoring data values and preserving the vari-
gram. By varying the starting point of random number generation, an ana-
st can play the game to determine how many high and low values may
ave been missed in the initial sampling, given the resultant variogram. The
)llowing images show the influence of changing the initial random number
gorithm on conditional simulation:

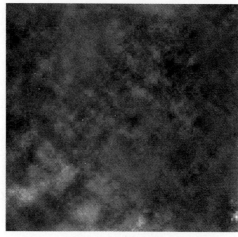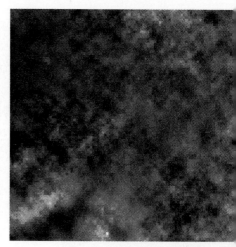

RNG started at position 1   RNG started at position 10001

There are similarities in these outcomes, for instance the influence of topography in the lower, left portion of the simulations. But, the visual appearance differs for these simulations in the lower, right corner. This illustrates how higher and lower values can shift in conditional simulation, even though the same data set is used for conditioning. This helps to determine how many higher and lower data zones may have been missed in initial sampling.

7.9 How Do I Reproduce Results in This Chapter, or Experiment With My Own Data . . .

7.9.1 . . . Using *Visual_Data?*

When using *Visual_Data* to generate **nonconditional simulations,** no data set needs to be opened in the main program window, moreover no data set need to be created, except when Hermite transformation is to be applied. In this case, an analyst should first open a data file in the main program window from which coefficients of the Hermite polynomial expansion will be computed. Once this file is opened, the analyst clicks Tools, then Histogram Analysis. At the conclusion of histogram analysis, *Visual_Data* prompts a user to create a file storing the Hermite polynomial coefficients. It is this file that is required by *Visual_Data* when transforming a nonconditional simulation.

The following steps are followed to obtain a nonconditional simulation:

1. click Tools, then choose Geostatistical Simulation
2. *Visual_Data* first asks a user if an earlier simulation is to be imported; this is useful if an untransformed simulation exists and a user wishes to apply a transform, or condition a nonconditional simulation. A user simply responds yes or no to this question;
3. *Visual_Data* then asks if the simulation is to be conditioned; the answer is no if the goal is nonconditional simulation; if the answer is yes, the program will inquire if the file containing the spatial data to be used for conditioning has been opened;

4. if the answer to step 2 is no, *Visual_Data* prompts a user to select the simulation algorithm: Gaussian (fractal) or turning bands (random lines);

5. a. **fractal:** a user specifies the desired size of the simulation (square), the starting point of the random number generation, an integer (whole) number from one to infinity, and the value of the Hurst dimension;

   b. **random lines/turning bands:** a user specifies the desired size of the simulation (square), the starting point of the random number generation, the number of lines to be used in the algorithm, often between 100 and 1000, and the range of the variogram; the range is specified in integer units relative to the size of the simulation; for example, a 100×100 simulation is developed that represents the entire area in which a spatial phenomenon occurs; this phenomenon has a range that is 40% of this size; a user specifies a range equal to 40 in this case (0.4 * 100).

6. once the simulation is completed, a user is prompted to save the simulation as a data file;

7. then, a user is also given the option to create a digital image of the simulation.

sing *Visual_Data* to obtain a nonconditional simulation is seen to involve a latively few steps, and it is reasonably fast.

Conditioning the simulation can take considerably more time. This ocess involves a double application of kriging, one application to the conditioning data, and a second application to simulated values closest to these cations. The larger the number of conditioning data, and the larger the mulation grid, the more time this step will take.

Conditioning proceeds as follows:

1. a file is created storing parameters necessary to the conditioning step; this file consists of three lines as follows:
   1. *Ymax, Xmin, Ydim, Xdim*
   2. *Co, Sill, Range, Anis, Ratio*
   3. *Jdat,* Ix, Iy, *Idat*
   defined as follows:

   > *Ymax, Xmin* are the *y* and *x* coordinates respectively of the upper, left corner of the grid created in nonconditional simulation; these coordinates must be consistent with the coordinates assigned to the conditioning data, *Ydim, Xdim* are the *y* and *x* spacings between grid rows and columns. *Co, Sill,* and *Range* are the nugget, sill and range, respectively, of the variogram for the conditioning data *Anis, Ratio* are parameters defining anisotropic properties of the conditioning data;

   > *Anis* = 0.0, *Ratio* = 1. for isotropic modeling;

   > *Jdat* is the number of data values on each line of the file containing the conditioning data. *Ix, Iy,* and *Idat* are the **positions** on each line of the *x* coordinate, *y* coordinate, and datum respectively.

As an illustrative example, the file used to condition the simulation of the visible blue reflectance for the *Nevada_Landsat_6x* data is as
follows:

$$0.5 \quad 0.5 \quad -1. \quad 1.$$

$$0. \quad 592.8 \quad 240. \quad 0. \quad 1.$$

$$9 \quad 9 \quad 8 \quad 1$$

Notice that Ydim is expressed as a negative number; why? Kriging expects the maximum y coordinate to occur at the top of a grid;
with digital images, however, row 1 ($y = 1$) is at the top of the grid,
and row, nrow ($Y = nrow$) is at the bottom of the grid. In the kriging
algorithm, the y-coordinate of each grid row is computed as: $Ystart$ -
i * Ydim, where i is the row number. $Ystart$ is equal to $Ymax + 0.5$
Ydim (this provides a half-grid cell offset for block kriging). To be
consistent with a digital image, $Ymax$ is set equal to 0.5, and $Ydim$
set equal to -1. Then, $Ystart$ is equal to zero, and the y-coordinate
each grid row is simply equal to $-i$ * Ydim.

This file is typed directly into the upper box of the main *Visual
Data* window; File is then clicked, and the option, Save, chosen
save this file; once saved, File is clicked, then Close is chosen, with a
questions responded to as No;

2. the file containing the spatial data for conditioning is opened in the
   main program window by clicking File, then Open;

3. the same seven steps as outlined for nonconditional simulation are
   now followed.

7.10   Literature

Chiles and Delfiner (1999) present a thorough review of all algorithms presently known for geostatistical simulation. Only two of the many algorithms are designed into *Visual_Data*. In part, these algorithms are chosen because they a
the best known to the author and because they are reasonably fast and easi
implemented using *Visual Basic*. Nonetheless, there are many other algorithm
and Chiles and Delfiner (1999) is recommended for review. Journel and Hu
jbregts (1978) is a recommended source when attempting the understanding
the turning bands algorithm. It is Chiles and Delfiner (1999) that inspired th
author to generalize this algorithm using any number of lines. The author
ability to effect this modification is the result of an initial understanding of th
turning bands algorithm obtained by reading Journel and Huijbregts (1978
Peitgen and Saupe (1988) is the recommended source for many numerical alg
rithms for simulating fractal processes. Pickover (1995) was inspirational fo
understanding how to develop high quality random number generators. E
periments presented in Section 7.2 on one-dimensional simulation are inspire
by Knudsen (1981), a notable tome on conditional simulation. Finally, reade
are referred to Mandelbrot (1983) and Hurst (1951) for background on, respe
tively, the fractal and Hurst dimensions.

xercises

. The program, RNG_FUN, located on the CD-ROM in the directory, \pcusers\programs\ allows a user to experiment with the design of random number generators. A general algorithm is provided of the form:

$$x = A(exp(x)) + Bx^C$$

A user manipulates the values of A, B, and C, in an attempt to design a high quality random number generator.

a. Design a generator of the form, $x = Bx^C$ by setting the value of A equal to zero.

b. Design a generator of the form, $x = A(exp(x))$, by setting the values of B and C equal to zero.

. The following Hermite coefficients are determined for a set of data. What are the transformed equivalents of:

a. -2.3 **b.** -1.4 **c.** 0.01 **d.** 2.0

Coefficients:	1	2	3	4	5
	5.81	1.005	−0.33	0.02	−0.02

3. Compute variograms for the visible green and near infrared (band 4) variables, *Nevada_Landsat_6x* data. Compute the Hermite polynomial coefficients for each of these variables. Develop several nonconditional simulations using the random lines (turning bands) and fractal algorithms. Determine Hurst dimension, H, from the variogram of each variable. Transform these simulations using both linear and Hermitian algorithms. Write a review of results.

4. Repeat problem 3 by varying the initial point of random number generation. Note especially changes in spatial patterns.

5. Condition the simulations obtained in problems 3 and 4 to the actual *Nevada_Landsat_6x* data. How do these visualizations compare to the actual data? Comment on their similarities, and especially their differences.

8

Digital Image Processing

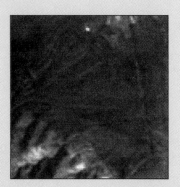

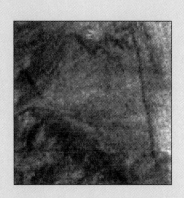

This text closes with two capstone chapters, the present one on digital image processing and a final chapter on data compositing. Digital image processing complements spatial data analysis through algorithmic manipulation, altering the visual appearance of these data thereby revealing patterns or features that perhaps are too subtle to discern in the unprocessed data. Given the widespread use of satellite-acquired and aircraft-acquired digital data, and digital photographs, an understanding of digital image processing is necessary to spatial data analysis. Digital image processing requires an understanding of basic statistics, multivariate data analysis, and geostatistics. Consequently, an overview of digital image processing provides a capstone summary of the text.

8.1 Pixels

The notion of the pixel is introduced briefly in Chapter 1. This notion is revisited to begin this chapter. A digital image is the same notion as square, or rectangular matrix of numbers. In the case of a satellite image, for instance, each of these numbers represents the average amount of reflected or emitted electromagnetic energy emanating from a square area. Each of these numbers is called a **pixel,** a connotation of value and area.

Pixel is a modern word developed from the phrase, **picture element.** Originally known as a **pel,** the term, pixel, has gained wider acceptance. In modern vocabulary, a pixel not only represents a digit making up a digital image, it also represents one of the many dots of color making up a computer monitor or color television screen. Depending on the capability of your computer and monitor—a typical computer graphics configuration—assuming MS Windows and Intel-based machines, allows 800 pixels across the screen, and 600 rows of pixels downward. The more pixels allowable across and down the screen, the more precise is the drawing of graphics on the screen. This is known as **screen resolution.**

A digital image is converted to a visualization by assigning each of its digits to a pixel location on a computer monitor screen. The numerical value of the image pixel governs the intensity at which the screen pixel is "turned on." For instance, if the image pixel value is 0 (zero), the screen pixel is black (no intensity). The larger the image pixel value is, the brighter the screen pixel is. If an image is displayed in black-and-white, as is the case with all previous displays in this text, the larger the image pixel value is, the lighter grey to white is its color.

This chapter focuses on pixels as **numbers.** Mathematical algorithms are applied to manipulate and change these numbers, consequently changing their visual context. Numbers have been manipulated throughout this text. In fact, pixels from a Landsat TM image have been used as the standard data et analyzed in each chapter. Pixels have also been created in many of these chapters. Examples include:

- new data coordinates (factors) based on eigenvectors; principal components transformation of digital images is demonstrated later in this chapter;
- estimates from kriging displayed as digital images;
- estimates from cokriging displayed as digital images;
- simulated data displayed as digital images.

This chapter more fully explores the many possible ways that pixels can be manipulated, and the visual consequences of these manipulations. The focus is on enhanced visualization of spatial phenomena afforded by these manipulations. Manipulations used in foregoing chapters, such as image differencing, are explained in more detail to reveal how some of these displays were created. This chapter further recalls many of the methods discussed throughout this text as a review to refresh and reinforce learning.

8.2   Adjusting Pixel Contrast

Previous chapters have presented actual Landsat TM images for comparison to analytical results, such as those from kriging, cokriging, and geostatistical simulation. These actual satellite images were adjusted for maximum pixel contrast to enhance their visual appearance, consequently rendering more obvious their information content. Why and how this is done is the partial focus of this chapter.

Given the visible blue reflectance of the *Nevada_Landsat_data* as one example, the following two displays illustrate the advantage of contrast maximization:

 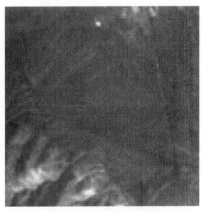

Original image and contrast   Enhanced image and contrast

Another way to visualize pixel contrast is by inspecting histograms of images. The histogram of the original image is compared to that for the enhanced image to show the numerical effect of contrast adjustment (shown on next page). Notice that the pixel values in the original image are associated with a relatively small range. One way to visualize contrast is the range. The larger the range, the greater pixel contrast is. Contrast is the same notion as pixel difference. Range is the difference, maximum pixel value minus

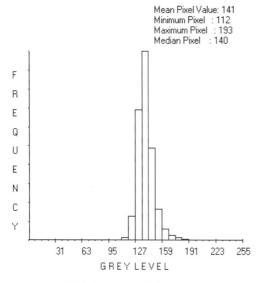

Original Image and Contrast

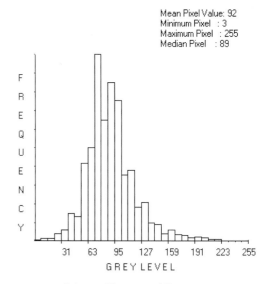

Enhanced Image and Contrast

minimum. Consequently, the larger the range is, the larger the difference between individual pixel values. The larger this difference is, the easier it to distinguish one pixel value from another. Recalling the theme of this tex visualization, the greater the pixel contrast, the better is image **clarity.**

8.2.1   Algorithms for Contrast Adjustment

Many algorithms are available for contrast adjustment. More thorough discu sion is presented in texts on image processing, such as Schowengerdt (1997 Only a few of the more common algorithms are discussed in this chapter.

To begin this discussion, a portion of a digital image is visualized. Se tion 8.1 equates digital images and matrices, and this concept is reinforced i this section by the following image segment, the upper, left 5 × 5 portion (the original visible blue image with original contrast, *Nevada_Landsat_data:*

140	138	137	136	137
138	137	139	139	137
140	140	138	139	141
139	136	134	131	135
137	134	132	132	133

Many satellite images yield, or are associated with pixel values writte as single-byte integers. A **byte** is a term that means computer word. A byte composed of eight bits, moreover a bit is a number, 0 or 1. Computers sto and manipulate numbers (information) using a sequence of electrical switc es. An electrical switch is either closed, allowing current to flow across it, open, blocking the flow of current. A closed switch represents the number, and an open switch represents the number, 0. Each switch is one bit. A sing image pixel requires eight switches for its representation.

Because two state electrical switches are used to represent information a computer, the base-two number system is used to convert such binary i

rmation to information that we humans are more accustomed to. For in-
ance, the integer value of a pixel is converted from its binary (switch, or bit)
quivalent as follows:

$$Pixel\ Value\ (Integer) = \sum_{i=0}^{7} b_{i+1}2^i$$

The bit, or switch value, b, is equal to 0 or 1. Its value "turns on or off"
e power 2 term.

EXAMPLE

In the 5×5 digital image segment shown above, one of the pixel values is 140. Its
binary representation is 10001100 (each 1 or 0 represents b_i, with i progressing
from right to left). This is confirmed using the following table:

i	7	6	5	4	3	2	1	0	
b_{i+1}	1	0	0	0	1	1	0	0	
2^i	128	64	32	16	8	4	2	1	
$(b_{i+1})(2^i)$	128 +	0 +	0 +	0 +	8 +	4 +	0 +	0 =	140

---

n understanding of the **radiometric resolution** of digital images, the maxi-
um numeric range of pixel values, is necessary to the correct interpretation
f their histogram. The foregoing two histograms indicate pixel values be-
veen 0 and 255, consistent with their 8-bit (single byte) radiometric resolu-
on. Some satellite data are transmitted using 16-bit radiometric resolution.
he greater the radiometric resolution is, the greater is the range of possible
ixel values, and the more precise is the measurement of electromagnetic en-
gy reflected or emitted by objects.

Recalling the 5×5 portion of the original visible blue image, several ex-
eriments are presented to adjust the contrast of these pixel values. Three al-
orithms are discussed explicitly:

- simple, linear;
- multiple, linear;
- histogram equalization.

2.1.1   Simple, Linear

ne of the standard approaches to contrast adjustment is the simple, linear
gorithm. Recall the histogram of the original visible blue image with origi-
al contrast. The minimum pixel value in the image is 112, whereas the max-
num value is 193 (these values pertain only to the upper 100×100
ortion of the original Landsat TM image). The range of these values is
93 − 112 = 81, significantly smaller than the range of 255 possible for 8-bit
teger quantities. Maximum contrast is achieved if we "stretch" the original
istogram range to utilize the full 255 range.

Simple, linear "stretching" is implemented as:

New Pixel = 255 × (Old Pixel − Minimum)/(Maximum − Minimum)

for which Minimum is that pixel value in the original image that will b reset to zero and Maximum is that value in the original image that will b reset to 255; then, all original pixel values are rescaled between 0 and 25 depending on their numerical relationship to the two end values, Minimum and Maximum.

EXAMPLE

Set Minimum = 112 and Maximum = 193 and rescale those pixel values show earlier for the upper 5 × 5 portion of the visible blue image. First, show one e ample calculation to rescale the original pixel value, 140:

New Pixel = 255 × (140 − 112)/(193 − 112) = 255 × (0.346) = 88

Contrast adjusted pixel values, upper 5 × 5 portion of the visible blue imag

88	81	78	75	78
81	78	85	85	78
88	88	81	85	91
85	75	69	59	72
78	69	62	62	66

Notice the effect of this contrast adjustment experiment. The first tw original pixel values, upper left, top row, are 140 and 138, with an origin contrast (difference) of 2. Upon application of the simple, linear transform the contrast between these first two values is 7 (88 − 81), more than a two fold increase.

Greater contrast is possible by using a smaller range, Maximum Minimum. Digital image processing is inherently experimental. There is a solutely no reason why a value, Minimum, needs to match the actual imag minimum value, and a value, Maximum, needs to match the actual imag maximum. For example, suppose in the foregoing example Minimum is ch sen to be 130 and Maximum is chosen to be 170. Then, the first two origin pixel values, upper left top row are transformed to:

New Value = 255 (140 − 130)/(170 − 130) = 63

New Value = 255 (138 − 130)/(170 − 130) = 51

The contrast between these two original pixels is now 12, a 6 fold increas from the original. Setting Minimum greater than the actual image minimur yields **saturation** in transformed pixels values equal to zero. All pixel value equal to, or less than Minimum are set equal to zero in the transforme image. Likewise, setting Maximum less than the actual image maximur yields saturation in transformed pixel values equal to 255. All pixel value equal to, or greater than Maximum are set equal to 255. Saturation, at the lo end, high end, or both, yields a harsher visual effect, but this may also e hance certain image features better than when no saturation is effected.

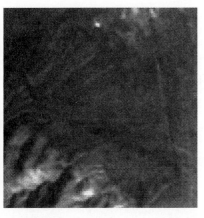 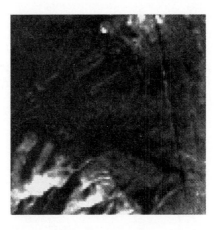

FIGURE 8.1

Two linear contrast stretches applied to the visible blue image. The image on the left is adjusted using Minimum = 112 and Maximum = 193. The image on the right is adjusted using Minimum = 130 and Maximum = 170, resulting in saturation of lower and higher pixel values. This results in a harsher outcome, darker lows and brighter highs, but certain features associated with midrange pixels are more distinguishable.

2.1.2   Multiple, Linear

This algorithm offers more flexibility (and experimental opportunity) in comparison to the simple, linear algorithm. Linear segments may be pieced together to model many different, complex stretching configurations. The simplest of these, a bilinear stretch, uses two linear segments. More than two segments enables differing enhancements for different ranges of pixel values.

In a bilinear enhancement, three values are specified to implement the stretch: x_1, x_2, x_3, defining the minimum, middle, and maximum values, respectively, for the stretch. For clarity, these values are illustrated on the right. A bilinear stretch is applied to the visible blue image, *Nevada_Landsat_data*, using $x_1 = 112$, $x_2 = 140$, and $x_3 = 192$:

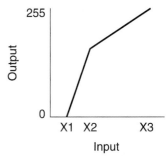

▲ FIGURE 8.2
Illustration of a bilinear stretch defined by two line segments: (x_1, x_2) and (x_2, x_3).

 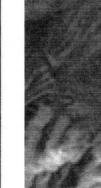

Simple, Linear   Bilinear

2.1.3   Histogram Equalization

This algorithm is nonlinear. The objective is to achieve more or less equal frequencies among the different pixel values, thus rendering a transformed image for which no particular range of pixel values is dominant. The visual outcome has a tendency to appear harsh.

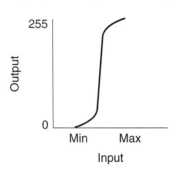

▲ FIGURE 8.3
Histogram equalization algorithm, in graphical form.

In this algorithm, illustrated in Figure 8.3, the cumulative frequency curve obtainable from the image histogram is used to implement the stretch. A demonstration is shown, applied to the visible blue image, *Nevada Landsat_data*. Moreover, the histogram before and after application of this algorithm is displayed to show the numerical result.

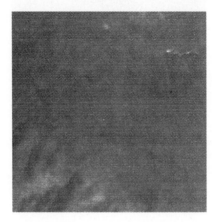

▲ FIGURE 8.4
Histogram equalization applied to the visible, blue image, *Nevada_Landsat* data. Left image: original; right image: after histogram equalization. Note harshness.

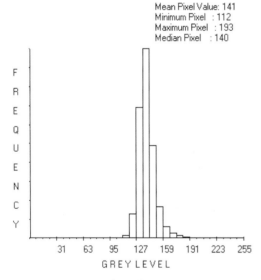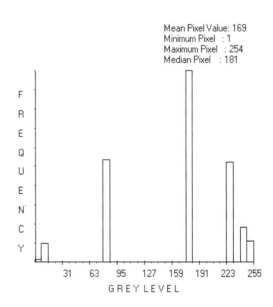

▲ FIGURE 8.5
Histograms before (left) and after (right) application of histogram equalization.

8.2.1.4 Binary Stretch

A useful stretch that can be implemented with the simple, linear algorithm is the binary stretch. In this type of contrast adjustment, the outcome is an image composed of pixels having only one of two possible numerical values, 0 or 255 (black or white). The objective of this type of stretch is the identification of the spatial region, or regions, of pixels of similar numerical value, and their spatial patterns.

For example, using the thermal image of the *Nevada_Landsat* data set, suppose we want to know the spatial locations of all pixels having values greater than 200 (higher thermal values). One way to effect a binary stretch is to use the simple, linear algorithm. Setting Minimum =Maximum = 200 will result in a binary outcome. In fact, several experiments are presented to obtain four binary stretches: 1) Min = Max = 50; 2) Min = Max = 100; 3) Min = Max = 150; and 4) Min = Max = 200. The binary stretching process is analogous to the indicator transform used in kriging:

A.   B.

C.   D.

▲ FIGURE 8.6
Four binary stretches of the thermal image, *Nevada_Landsat_data*. A. black shows region of pixels less than 50; B. black shows region of pixels less than 100; C. black shows region of pixels less than 150; and D. black shows region of pixels less than 200. Note how this experiment helps to focus attention on particular ranges of pixel values.

In another experiment with binary stretching, only those pixels that are between 150 and 200 are delineated (white) with all other pixel values displayed as black. A custom stretch is designed. In this experiment, the custom stretch is implemented by designing a *look-up table* for the stretch. In the computer implementation of contrast enhancement, numerical computations are only necessary to define 256 entries in a look-up table. These 256 entries are the output equivalents of the input pixel values. For instance, suppose a simple, linear stretch is desired using Min = 100 and Max = 200.

A look-up table is designed as a data set of 256 values, one value per line i⁻
the data file, as follows:

Line 1 = 0
Line 2 = 0
Lines 3 through 100 = 0
Line 101 = 255(100 − 100)/(200 − 100) = 0
Line 102 = 255(101 − 100)/(200 − 100) = 2
Lines 103 − 200 = 255 $(x − 100)/(200 − 100)$, for x = 102, 103, ..., 199
Line 201 = 255
Line 202 = 255
Line 203 − 256 = 255

Why is a look-up table necessary? Assume that an image exists that re⁻
quires contrast adjustment, and its size is 2000 rows × 2000 pixels per row. Thi⁻
image consists of 4 million total pixels. One approach to contrast adjustment o⁻
this image is to complete 4 million calculations, one for each pixel location t⁻
determine output pixel values. But, each of the 4 million pixels has a value tha⁻
ranges only between 0 and 255. Consequently, it is computationally more eff⁻
cient to make 256 calculations to compute the output equivalent for each of th⁻
256 possible input pixel values. This saves considerable computational effort a⁻
output values are obtained from the look-up table, not computed.

A look-up table is designed for the custom binary stretch as follows:

Lines 1 − 150 = 0
Lines 151 − 200 = 255
Lines 201 − 256 = 0

This custom look-up table is saved as a 256 line data file, one value pe⁻
line, then is applied to the thermal image. The outcome is shown below.

▶ FIGURE 8.7
Custom binary stretch to highlight the thermal
image, pixel region 150 − 200 (white), and all
other pixels values are shown in black.

8.3   Filtering Digital Images

Filtering is synonymous with removing. Particular information is remove⁻
from a digital image upon application of a filtering algorithm. Many algo⁻
rithms exist for filtering digital images, each with a different purpose, or in⁻
tended type of image. The following list presents a few examples:

- *low-pass filter:* removes high frequency image information, and preserves (passes) low frequency information; this filter is useful for attenuating high frequency noise; this type of filter smooths (blurs) an image and this can be a beneficial preprocessing step in digital image classification; low-pass filtering is mathematically analogous to kriging;

- *high-pass filter:* removes low frequency image information, and preserves (passes) high frequency information; this filter is useful for enhancing edges and lines; moreover, low-frequency noises are attenuated; this type of filter is usefully applied in geology for fault detection;

- *high-boost filter:* this filter does not remove any image information; instead, high frequency content is amplified; the visual consequence is a crispening, or sharpening of the image by making edges (boundaries) sharper;

- *median filter:* this filter removes random, spatially uncorrelated noise by replacing each pixel in an image by the median pixel value computed for the local region surrounding the pixel; the optical effect is similar to low pass filtering, but blur is possibly not as severe;

- *Olympic filter:* a specialized filter; pixel values in the original image are replaced by the average, (Max + Min)/2, where Max and Min are the maximum and minimum pixel values respectively for the local region surrounding the pixel; this filter derives its name from the manner in which athletes' scores are computed in the Olympics;

- *Sigma (Lee) filter:* this is a specialized filter, especially for application to microwave images to attenuate "speckle" noise; this filter is a form of low-pass filter; pixels in the original image are replaced by subtracting from their original value a factor, $k\sigma$, with $k\sigma$ equal to the standard deviation of the noise that is to be removed, and σ is the standard deviation of pixels in the local region surrounding a pixel to be filtered; k is an experimentally derived factor that represents the proportion of σ that is noise.

These are the filters that are specifically offered by *Visual_Data*. Additionally, image analysts may design custom filters for implementing many different filtering strategies, most notably spatially directional filters.

8.3.1   In a Digital Image, What Is a Low Frequency Feature? What Is a High Frequency Feature?

Certainly, if one is to understand spatial filtering completely an understanding of image frequency content is essential. A simple experiment helps when trying to understand what in an image is a low frequency feature or high frequency feature.

EXPERIMENT

Hold this text book about 60 cm (2 ft) away from your eye(s). Squint. Notice as you squint that individual words and text are no longer discernable. The shape of the page, its white color, and a grey blur in the center due to the black text are discernable. As you squint, focus is lost. Despite lost focus, shapes and colors remain discernable. A similar experiment is possible using a camera or telescope; when out of focus, shapes and colors are still discernable. What is no longer discernable when you squint is the finer detail, the individual letters on the page are lost.

> Gross detail, the *shape* of the textbook page, its *color (tone),* are examples of *low frequency* features. Low frequency features are those that change slowly over space. Fine detail, the lines that distinguish the letter, "a," from the letter, "e," for instance, the edges of a crater when looking at Earth's moon through a telescope, are examples of *high-frequency* features. High frequency features are those that change rapidly in space. The sudden transition from one rock type to another, the *boundary* or *edge,* is another example of a high frequency feature.

8.3.2 Spatial Convolution Filtering: Low-Pass, High-Pass, High-Boost, and Custom Strategies

Weighted averaging is used to implement these filtering strategies. Theoretically, a convolution is solved as an integral:

$$Z(x) = w(x) * O(x) = \int w(x)O(x)dx$$

with O representing the original image, w representing the filter, Z representing the filtered image, and * representing the convolution operation. In practice, the integral is solved as a summation:

$$Z(x) = w(x) * O(x) \approx \sum_{i=1}^{N} \sum_{j=1}^{N} w_{ij}(x)O(x)$$

In fact, this may be written using the exact formula that is used for kriging

$$Z^*(x_0) = \sum_{i=1}^{M} \lambda_i Z(x_i)$$

where, in this case, M is used as the limit of summation and is equal to N. The value, N, represents the size of the filter, a square configuration of weights. The larger N is, the more severe is the effect of the filter. The concept of N, applied to the filtering of an image, is sketched in Figure 8.8.

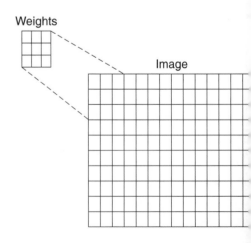

▶ FIGURE 8.8
A square configuration of weights applied to the filtering of a digital image.

This figure shows that the regular geometry of a digital image is taken advantage of in the design of the filter. The geometric configuration of pixels is the same for each calculation of an output pixel. Consequently, the weights are invariant with spatial translation. In comparison, the geometric configuration of nearest neighboring locations used in kriging often changes as the estimation location changes and weights are necessarily computed for each

ew geometric configuration. Except for this distinction, spatial convolution
ltering and kriging are the same concept.

Several examples illustrate spatial convolution filtering. A simple low-
ass filter is obtained using the following filtering strategy:

1/9	1/9	1/9
1/9	1/9	1/9
1/9	1/9	1/9

This strategy uses $N = 3$; each weight is equal to $1/N^2$. A low-pass filter
s implemented in this case as a simple average of the pixels within a 3×3
vindow. This is but one example. In general, a weighting strategy for which
he following is true, $\Sigma \lambda = 1$, often results in a low-pass filter. Recall that
his weighting scheme is also that which is used in kriging.

A visual example of low-pass filtering is necessary for a complete under-
tanding. In keeping with examples from previous chapters, the same, small
ubsampling of a Landsat TM image is used for this demonstration. Two fil-
er sizes, $N = 3$ and $N = 7$ are used to show the influence of filter size on the
isual outcome:

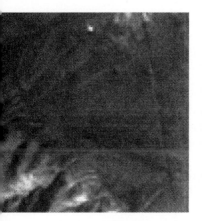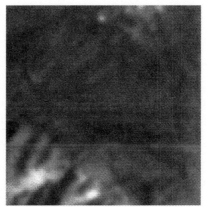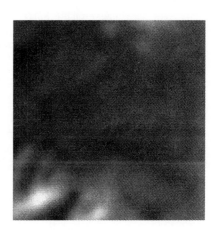

▲ FIGURE 8.9
Two low-pass filters applied to the visible blue image, *Nevada_Landsat* data. The image on the
left is the original image; the center image shows a 3×3 low-pass filter; note the blur but
some linear features are still evident. The right most image is a 7×7 low-pass filter. Severe
blurring is noted and most high frequency content is removed.

Only the weighting matrix is changed in spatial convolution filtering to
ffect a different filtering strategy. For example, the following table of
veights is used to implement a high-pass filter:

−1/9	−1/9	−1/9
−1/9	8/9	−1/9
−1/9	−1/9	−1/9

ι this filtering strategy, the weights sum to zero. Recall the method of cok-
ging, in which a primary variable is estimated by applying weights to it
hat sum to one, and applying weights to auxiliary information that sum to
ero. It is stated in Chapter 6 that this weighting strategy applied to auxiliary
nformation adds high frequency information to the primary variable, pro-
ided the resolution of the auxiliary information is higher than that of the
rimary variable. This statement is based on the observation that weights

applied to the auxiliary information sum to zero. In general, a weightin
scheme that has a zero sum often implements a high pass filter.

Visualizations are quite important for underscoring the optical effect
high-pass filtering. Two filter sizes are used, $N = 3$ and $N = 7$ to show th
effect of filter size on the enhancement of high frequency image informatio

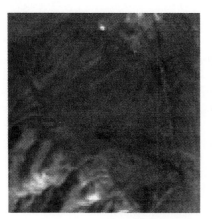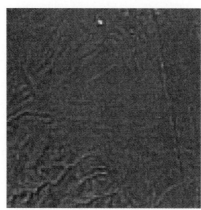

▲ FIGURE 8.10
Two high-pass filters applied to the visible blue image, *Nevada_Landsat_data*. Left most image
is the original. The middle image is a 3 × 3 high pass filter. The right most image is a 7 × 7
high pass filter. Line detail is enhanced by a high-pass filter; lines, or edges in general, are high
frequency image features. Tones (colors) are low-frequency features. Notice that these high-
pass filters have removed tones (darker and lighter shades).

High pass filters are used when edge and line detail are more important t
an image analyst than tone or color information. In geology, high-pass filte
ing is especially useful for structural geology studies, particularly whe
looking for faults, joints and folds.

By changing the high-pass filter weighting scheme slightly, a high-boo
filter is obtained. As is stated previously, this method does not remove imag
information. Instead, high-frequency information, such as lines and edges,
amplified in numerical importance. The visual effect is a sharpening of th
image, as is demonstrated in the following figure.

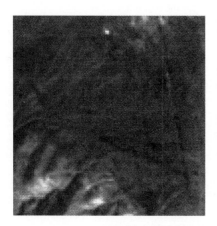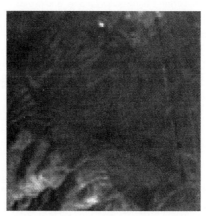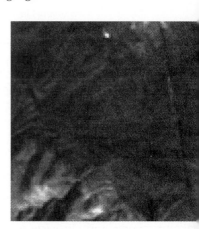

▲ FIGURE 8.11
High-boost filtering applied to the visible blue image, *Nevada_Landsat_data*. The image on the
left is the original. The middle and right images are both 3 × 3 filters. The middle image is
based on a boosting factor of four, whereas the image on the right is based on a boosting factor of eight. The greater the boost, the crisper is the outcome.

This figure compares optical results from two levels of boost. The following two tables show the weighting schemes that were used to obtain these results.

8.3.2.1   Boosting Factor of 4:

−1	−1	−1
−1	32	−1
−1	−1	−1

8.3.2.2   Boosting factor of 8:

−1	−1	−1
−1	64	−1
−1	−1	−1

The center weight is the key factor that is manipulated in high-boost filtering. Weights in high-boost filtering sum to a value that is greater than one. The larger this sum is, the more severe is the effect of the filter.

Spatial convolution filtering is inherently experimental. Custom design of weighting schemes is therefore an essential skill for image analysts. One example is the design of **directional,** high-pass filters. Some image features are high-frequency phenomena only in a particular direction; in another direction they are low-frequency phenomena. A highway is one example. This human-made phenomenon is a **low frequency** feature in the direction of vehicle traffic, but is a **high frequency** feature perpendicular to the flow of traffic. A natural example analogous to this is a fault. A fault is a low-frequency feature in the direction of its **strike,** but is a high frequency feature in the direction of its **dip.**

Two 3 × 3 custom high-pass filters are designed, one that enhances the east-west direction, and the other that enhances the north-south direction.

8.3.2.3   East-West Custom High-Pass Filter (3 × 3):

0	0	0
−1	2	−1
0	0	0

8.3.2.4   North-South Custom High-Pass Filter (3 × 3):

0	−1	0
0	2	0
0	−1	0

Notice that the weights in spatial convolution filtering are "mini" images. Recall the definition of a digital image forwarded at the beginning of this chapter, a matrix of numbers. Each of these weighting matrices is consequently a digital image. Spatial convolution filtering is therefore the convolution (multiplication) of one image by another. Direction in each image is the same as that on a map, the 360 degree circle directions that we are accustomed to. Designing directional weighting schemes is based on these 360 degree directions. Notice the east-west and north-south examples. Weighting is only in the horizontal (0; 180) direction in the east-west design, and only in the vertical (90; 270) direction in the north-south design.

These custom weighting schemes are applied to the visible blue image *Nevada_Landsat_data*. The intent is to see what features are enhanced by each feature, but more importantly what features are lost. By this inspection, the understanding of image frequency content is reinforced.

 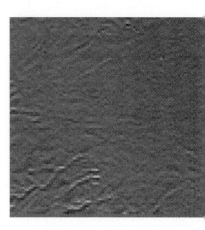

East-West High-Pass   North-South High-Pass

Do you notice the differences in these results? Note especially the roadway on the right side of the east-west enhancement. This roadway (linear feature) has a trend (strike) in the northwest-southeast direction. This is a low frequency feature in this strike direction, but is a high frequency feature in a direction more or less perpendicular to this direction. The east-west direction is approximately orthogonal to this linear feature and it is consequently enhanced well by the east-west directional filter. In contrast, because this linear feature is a low frequency phenomenon in the north-south direction, it is filtered (removed) by the application of a filter in this direction. Note this linear feature is not visible in the north-south enhancement.

8.4   Principal Components Analysis of Multispectral Digital Images

Principal components analysis of multivariate data is the focus of Chapter 4. In the context of digital image processing, eigenvectors are visualized as digital images by projecting original image pixels onto them. Each projection is known as a **principal components** image. Moreover, each principal components image is a function of *all* spectral images, not just one. These projections offer unique opportunities for visualizing multispectral image data.

Projecting pixel values onto eigenvectors is mathematically more simple than the computation of factors used in Chapter 4. In the case of digital image processing, principal components analysis proceeds as follows:

Step 1:   Let the entire collection of multispectral image pixels be called $[Y]$ an $N \times M$ matrix; M is the number of spectral bands, seven in the case of Landsat TM data; N is the total number of pixels in each spectral band, for example, if an image consists of 100 rows each with 100 pixels, then N is equal to 100×100, or 10000;

Step 2: Compute a matrix, $[S]$, $M \times M$ in size; recall Chapter 4 and the notion of this matrix; $[S]$ is a variance/covariance matrix for principal components analysis; or, $[S]$ is a matrix of correlation coefficients for standardized principal components analysis; or, $[S]$ is a matrix of chi-square distances for correspondence analysis;

Step 3: Decompose $[S]$ into eigenvalues and eigenvectors;

Step 4: Compute principal components images, $[P]$, as the product, $[Y][E]$, in which the columns of the matrix, $[E]$, are the individual eigenvectors of $[S]$.

Step 5: Display the individual principal components images for analysis.

Visual examples are essential to a complete understanding. In this example, all seven bands of the *Nevada_Landsat* data set are treated simultaneously to obtain the principal components images. All three algorithms, principal components analysis, standardized principal components analysis, and correspondence analysis are applied for comparison and contrast.

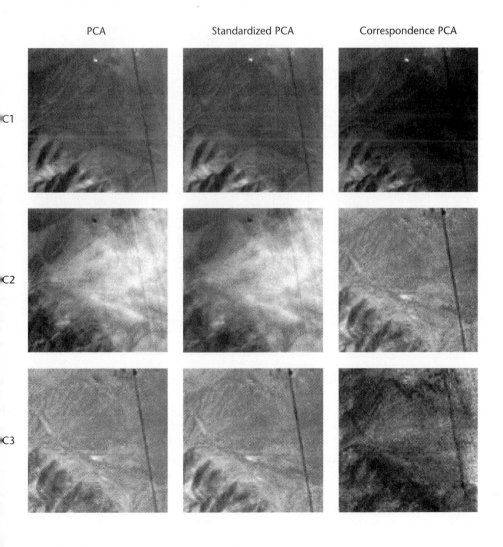

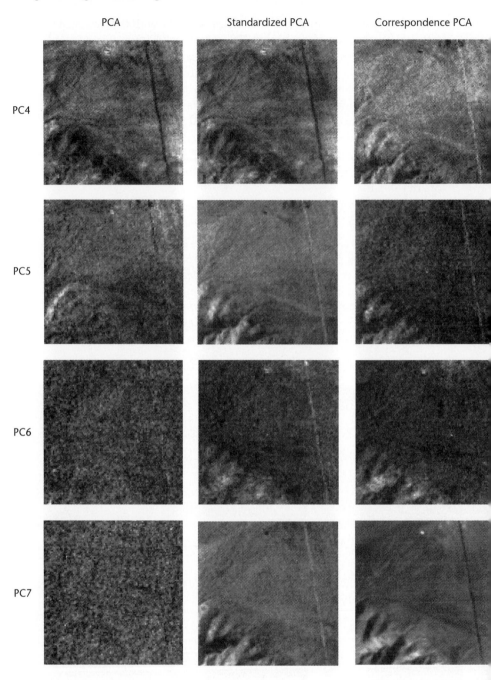

This is an interesting outcome in terms of algorithmic performance. Although PCA and Standardized PCA yield practically identical results for the first several principal components images, they yield distinctly different results for the last several PC images. Correspondence analysis yields results that differ from PCA and Standardized PCA for the same PC image PC1, PC2, and so on. There is some similarity in algorithmic performance between correspondence analysis and the other two methods across PC images. Notice, for example, that the fourth PC image from correspondence analysis is similar to the fifth PC image from PCA and Standardized PCA.

Focusing solely on principal components analysis (PCA), notice that "noise" becomes more obvious with each PC image. The first PC image is mostly signal. Ground features are clear and easily distinguishable. In contrast, the seventh PC image is mostly noise. Ground features are no longer distinguishable. This result is typical when using the PCA algorithm.

The other two algorithms yield markedly different results for the seventh PC image. In fact, correspondence analysis yields a seventh PC image that is mostly signal. Standardized PCA also yields a seventh PC image that is mostly signal. Explaining this outcome requires examination of the eigenvalues obtained from eigendecomposition. These values are presented in the following table.

Eigenvalue	PCA	%	SPCA	%	CA	%
1	414.9	78.1	5.50	78.6	9.66E-04	63.1
2	69.4	13.1	1.00	14.3	3.11E-04	20.3
3	32.7	6.2	0.30	4.3	1.73E-04	11.4
4	7.6	1.4	0.11	1.5	5.28E-05	3.5
5	4.7	0.9	0.06	0.9	1.52E-05	1.0
6	0.9	0.2	0.02	0.3	1.15E-05	0.7
7	0.7	0.1	0.01	0.1	−2.90E-10	0*
Total	530.9	100	7	100	1.53E-03	100

*Negligible
PCA is principal components analysis   SPCA is standardized PCA   CA is correspondence analysis

The focus is on results from correspondence analysis (CA). Notice for the seventh eigenvalue that the relative contribution from the seventh principal component is negligible. The seventh eigenvalue is five orders of magnitude smaller than the sixth. When eigenvalues approach zero, the corresponding eigenvector approaches unity. The seventh eigenvector from correspondence analysis is, in fact,

$$\{1.24 \quad 0.87 \quad 1.03 \quad 0.95 \quad 1.31 \quad 1.44 \quad 1.00\}.$$

All entries in the vector are not precisely equal to one, but are roughly equal to one. Projection of original pixels onto such a vector yields approximately the original pixel values, and this explains the visual outcome from correspondence analysis for the seventh PC image. This seventh PC image is negligible, judging by the size of its associated eigenvalue.

Correspondence analysis always yields a reduction in original variables by at least one. Carr and Matanawi (1999) cited this fact for possible application to digital image compression. Compression is a concept often discussed for improved efficiency in data storage. If data can be compressed, less storage space is required. The price for this improved efficiency may be a loss of original data integrity. Correspondence analysis yields improved data reduction in comparison to PCA and SPCA without a loss of data integrity.

Switching the focus to results from SPCA, the importance of its seventh principal component is approximately that obtained using PCA. Both algorithms are associated with seventh principal components that account for approximately 0.1% of the original image information. The key word in this sentence is "approximately."The seventh principal component in SPCA actually accounts for 0.14%, in comparison to 0.13% for PCA. This seems trivial, but this distinction is substantive with respect to the eigenvector:

$$\text{PCA:} \quad \{3.4 \quad -16.5 \quad 10.1 \quad -3.3 \quad -0.6 \quad -0.8 \quad 1.0\}$$

$$\text{SPCA:} \quad \{0.1 \quad 2.4 \quad -6.0 \quad 3.7 \quad -1.2 \quad 0.5 \quad 1.0\}$$

This is a considerable numerical difference and helps to explain the different visual results from these algorithms for the seventh PC image. This state-ment is underscored by comparing the eigenvectors from these two algo-rithms for the first PC image for which results are visually more similar:

$$\text{PCA:} \quad \{1.2 \quad 0.7 \quad 1.0 \quad 0.8 \quad 1.4 \quad 0.5 \quad 1.0\}$$

$$\text{SPCA:} \quad \{1.0 \quad 1.0 \quad 1.0 \quad 1.0 \quad 1.0 \quad 0.4 \quad 1.0\}$$

These eigenvectors are quite similar, and so are the consequential PC images. In summary, very small differences in percent importance of principal com-ponents may have sizeable impact on visual appearance of the PC images, as is observed in the seventh PC images from these two algorithms.

8.5   Dust Devils on Mars

Image processing is the visual translation of spatial data. Foregoing exam-ples illustrate the type of informaiton that can be enhanced, or eliminated through image processing. This section illustrates how features in digital images that are quite subtle may be enhanced and rendered substantially more obvious.

Dust Devils, a common atmospheric phenomenon on Earth, were first identified on Mars in 1985 (Thomas and Gierasch, 1985). Their analyses were based on digital images returned by the Viking orbiters in 1976.

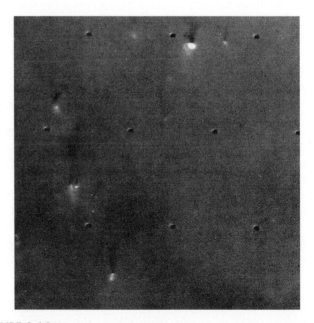

▲ FIGURE 8.12
A portion of Viking image, 034B01. Dust devils appear as bright [white] features that cast shadows toward the top of the image.

The Viking mission involved two soft landings on Mars. Each lander system included a digital imaging camera. Given the mechanics of the camera system, a relatively slow mechanical operation, dust devils were not observable in images returned by these landers. Dust devils are dynamic, and relatively fast camera operation is necessary to their successful capture on film or digital media.

In July, 1997, Mars Pathfinder was the first United States mission to return to Mars with a surface lander since the 1976 Viking program. Pathfinder included a digital imaging camera capable of shutter speeds sufficient for capturing rapidly moving dust devils. Moreover, the digital imaging system, known as IMP (Imager for Mars Pathfinder) had a spectral resolution that included separate filters, visible blue, green, and red. This enabled the compositing of these filters into true color images of the Martian surface and horizon.

Additionally, Mars Pathfinder included a weather mast. Temperature, barometric pressure, and wind speed was monitored. This instrument detected a few atmospheric phenomena that included rapid barometric pressure and wind speed fluctuation, as well as brief temperature fluctuation (Schofield, et. al., 1997). Analysis of these data indicated the passage of a dust devil over the Pathfinder lander as the most likely cause. Yet, no true color composite of images from Pathfinder revealed an obvious dust devil.

Metzger and Carr (1996) speculated that a specialized image processing technique is necessary to the surface identification of dust devils. Viking landers, although lacking the imaging capability to clearly identify dust devils, nonetheless yielded images that showed a heavily dust laden lower Martian atmosphere. Given the reddish hue of the Martian soil, a spectral characteristic attributable to the oxidation of iron minerals, the lower Martian sky also has a reddish hue given its dust loading. A reddish Martian dust devil consequently has a spectral signature quite similar to background Martian sky. A dust devil is thought to contain more dust than background sky, therefore its red signature should be somewhat higher. By subtracting a blue filter image from a red filter image of the same horizon, the red signature of the dust devil should be boosted relative to background sky. This image differencing was thought to be necessary to dust devil identification in Mars Pathfinder images.

8.5.1   True Color Compositing: The 24-Bit Bitmap

Microsoft Corporation developed a particular file format for digital images to optimize their handling and display by the Windows operating system. This format was developed primarily for images of icons and symbols, yet digital images of any type are handled well by Windows if in this format. Known as the Bitmap (.BMP) format, this format for digital images is universally accessible by all Windows based software programs.

There are several advantages to converting digital image data to BMP format. Windows is able to display such files rapidly. **A header record,** information at the beginning of the image data, defines the size of the image (number of rows and pixels per row), and the type of bitmap, single 256 color image, or three-image, 24-bit composite. Of particular use in educational circumstances, the Windows operating system automatically adjusts the color content of these bitmaps to match the capability of a user's hardware. For instance, a 24-bit bitmap consisting of three 8-bit images, can be displayed with a maximum of 2^{24} (16,777,216) colors. If a user's computer has a maximum color capability of 16-bit color (65,536 colors), the Windows operating environment automatically adjusts the 24-bit display to 16-bit. Optimum system performance is therefore achieved when using digital images that are written in BMP format.

Visual_Data has a tool for converting raw image data to Windows BMP format. This tool requires a user to know the size of the image, number of rows and pixels per row, as well as the number of header bytes, if any, at the beginning of the image file. If image pixels are written as 8-bit integers, one byte per pixel, the number of header bytes can be determined by multiplying the number of rows by the number of pixels per row (total image bytes), and subtracting this product from the total number of bytes for the file. Often, the number of header bytes is zero. If a user is analyzing planetary data written in PDS (Planetary Data System) format, a header record is included in the file. This header is described in a separate label (.LBL) file that is written in ASCII format, a common text format. Opening the file in a program such as Word pad (Windows: Programs: Accessories), should allow a user to read the head er record. Therein, the number of rows and pixels per row are documented.

To begin the image processing of the Mars Pathfinder images, digital im ages from the individual color filters, blue, green, and red, are first converted to Windows bitmap format. Mars Pathfinder images consist of 248 rows, each with 256 pixels. These pixels are written in 16-bit format, higher numerical precision (radiometric resolution) than 8-bit pixels. Example images are included on the CD-ROM in the directory, \images\marspath\. Two Martian locations are cho sen as examples, the southern edge of Twin Peaks and a view across Big Crater toward Misty Mountain. Analysis of Twin Peaks images is left as a challenge to students at the end of this chapter. The focus is now on Misty Mountain.

As noted earlier, dust devils are not obvious in true color composites of the Mars Pathfinder images. The Misty Mountain red, green, and blue filter images are converted to Windows bitmaps, then composited into a single 24-bit bitmap using *Visual_Data*. The result is displayed as a black-and-white rendition (Figure 8.13). Careful examination of this image does not result in the identification of one, or more, dust devils.

▲ FIGURE 8.13
True color composite of Misty Mountain (small knoll in the distance). Its "misty" appearance is attributable to dust loading in the lower Martian atmosphere. Dust devils
are not obvious in this true color display. (See CD-ROM, folder\colrfigs\ for color ver-

An experiment is attempted in which the blue filter image is subtracted from the red filter image of Misty Mountain. An alternative is to compute a ratio of images, red filter divided by blue filter. Both mathematical operations yield similar optical results (Figure 8.14). Both results are contrast adjusted for maximum optical display. Particularly noteworthy is the bi-directional cross-hatching noise. This artifact is attributable to projection of light through a convex lens onto a flat charge coupled device (CCD). The cross-hatch noise is the result of manufacture variation in the surface of the CCD. The original proposal by Metzger and Carr (1996) did not anticipate noises attributable to camera performance.

▲ FIGURE 8.14
Image difference, left, red filter minus blue filter. Image ratio, right, red filter (numerator image) divided by the blue filter (denominator image). Note the cross-hatch noise. These two approaches to image processing do not clearly identify dust devils.

A common image processing technique used in planetary study is the subtraction of a "flat-field" image from a "target" image as a prelude to other forms of image processing. A flat-field image contains system artifacts that are to be removed. In the case of the Mars Pathfinder images, for example, a flat-field image is contaminated by the same camera noise, such as the CCD cross-hatch noise, that contaminates other images. Incidentally, this noise varies with filter. In other words, the geometric pattern of cross-hatching is different, for instance, for the blue filter than it is for the red filter. Each color filter must be corrected by a flat-field image for that filter.

In the case of the Mars Pathfinder images, because the objective is the identification of dust devils, "sky-flat-field" images are sought. These images show only a minimal amount of ground surface in their lower portion, otherwise they are mostly sky. Dust devils are most easily identified in images of horizon because they are silhouetted against background sky. Using sky-flat-field images to correct for camera artifacts imparts only a minimal artifact at the bottom of each processed image that is due to the subtraction of the sky-flat terrain geometry from that of the target image.

An experiment involves the subtraction of a blue filter sky flat from the blue filter image of Misty Mountain. The image difference is contrast enhanced for maximum effect and displayed in Figure 8.15.

▲ FIGURE 8.15
Misty Mountain: blue filter image minus the sky flat blue filter image. Two dust devils are identified in this enhancement, the largest and most prominent of which is on the right side of this image. The smaller dust devil appears to be on the top of the Misty Mountain knoll. Images: original blue filter (left), sky flat (middle), and processed image showing dust devil features (right).

Notice in Figure 8.15 that the subtraction of a blue filter sky flat image from the Misty Mountain blue filter target image has eliminated (filtered) the cross-hatch artifact attributable to the camera's CCD. This cross-hatch noise is spatially variable. Consequently, a Fourier domain filtering approach will not eliminate it. The sky flat subtraction approach is necessary to the elimination of this noise.

Cross-hatch noise varies with spectral frequency for the Mars Pathfinder images. Consequently, sky flats for the same frequency band must be used to correct for this noise. Using, for instance, a red-filter sky flat to correct a blue filter image will not work. This is underscored visually in Figure 8.16.

A serendipitous discovery was made when applying principal components transformation to 24-bit true-color bitmaps of Mars Pathfinder scenes known to be associated with dust devils. This discussion involves the application of correspondence analysis for the transformation, but a similar result is obtained when using principal components or standardized principal components analysis. In correspondence analysis, given that three images are transformed, two substantive transformations are obtained; the third PC image is discarded as representing no useful information. The first two principal components images of Misty Mountain are shown in Figure 8.17 after contrast adjustment. Correspondence analysis (principal components transformation) has identified a dust devil feature in the second PC image of the Misty Mountain scene without the sky flat subtraction preprocessing step.

▲ FIGURE 8.16
An experiment that involves the subtraction of a red-filter sky flat from the blue filter Misty Mountain image. Notice that the cross-hatch noise is not removed. The spatial nature of the cross-hatch noise is different for the red-filter image than for the blue filter. In this case, the dust devil on the right side of this image is vaguely evident at the horizon, but this is a realization that is partially inferred from Figure 8.15.

◀ FIGURE 8.17
Correspondence analysis transformation of the Misty Mountain true color composite image. The first (left) and second (right) principal components images are displayed. A tri-linear contrast adjustment was applied to the first PC image. A histogram equalization was applied to the second PC image for visual effect given the black-and-white display on this page. Although quite harsh, notice the identification of the dust devil feature in the second PC image.

8.6   How Do I Reproduce Results in this Chapter, or Analyze My Own Images . . .

8.6.1   . . . Using *Visual_Data?*

Visual_Data is designed to facilitate image processing. This feature is one of the most interactive aspects of the program. A user proceeds directly to Tools once starting the program. There is no need to first open a file in the main program window, unless custom spatial convolution filtering is the intended goal. In this case, the data file containing the weights for the custom filter is opened in the main program window before proceeding to the image processing tools.

Digital images must be in the Windows BMP format to be compatible with *Visual_Data*. Often, digital images are not in this format. Some suggestions are forwarded for handling various digital image formats:

- digital images downloaded from the Internet are often in JPEG (jpg) or GIF format; the Windows accessories program, PAINT, should be able to convert these formats to BMP format. Click Start, then choose Programs, then accessories, then Paint, click File, then Open, find the digital image file to be converted and double click on it; Paint should start the file format conversion automatically; once the image is displayed, click File, then click Save As; set the file type to 24-bit bitmap for a color image, or 256 bitmap if the image is a single plane, 8-bit black-and-white image;
- digital cameras often store images in JPEG (jpg) format; some cameras are supplied with software for image file format conversion; otherwise, the Windows Paint program can be used to convert these image file formats;
- many digital images are written as raw pixel values, bit by bit, in a binary file, with or without a **header record.** These records are binary or text information written at the start of a digital image file containing information about the image, particularly image size in terms of number of rows of pixels and number of pixels per row; if this header is written as text information, the digital image file can be opened in any text editing program, such as the Windows accessories program, Wordpad, and information about the image can be obtained.

EXAMPLE

Digital images from NASA (National Aeronautics and Space Administration) are often written in PDS (Planetary Data System) format. This format consists of a header record, N bytes long, followed by raw pixel values, written bit by bit in binary form. The header is written as binary information, consequently it is NOT accessible using the Windows program, Wordpad.

Instead, separate files with .LBL extensions (short for "label") are supplied. These are text files that contain information about the images. An example shows the label file for one of the Magellan radar images of Venus (CD-ROM, directory \images\Venus\):

```
CCSD3ZF0000100000001NJPL3IF0PDS200000001 = SFDU_LABEL
/* Framelet file format, size and location */
RECORD_TYPE= FIXED_LENGTH
RECORD_BYTES  = 1024
FILE_RECORDS  = 1026
^IMAGE_HEADER = ("FF01.IMG",1)
^IMAGE  = ("FF01.IMG",3)
/* Framelet description */
DATA_SET_ID = 'MGN-V-RDRS-5-MIDR-FULL-RES-V1.0'
SPACECRAFT_NAME   = MAGELLAN
MISSION_PHASE_NAME     = PRIMARY_MISSION
TARGET_NAME= VENUS
IMAGE_ID   = 'F-MIDR.10N076;1'
INSTRUMENT_NAME   = 'RADAR SYSTEM'
/* Description of objects contained in the framelet */
OBJECT  = IMAGE_HEADER
TYPE = VICAR2
BYTES   = 1024
```

```
RECORDS = 2
END_OBJECT
OBJECT  = IMAGE
LINES   = 1024
LINE_SAMPLES   = 1024
SAMPLE_TYPE= UNSIGNED_INTEGER
SAMPLE_BITS= 8
NOTE = "
DN   = INT((MIN(MAX(RV,-20),30) + 20) * 5) + 1,
where RV   = radar crossection/area divided by the
Muhleman Law and converted to decibels. Muhleman Law
multiplicative constant of 0.0118 was used. (Note: Intention
was to use 0.0188.)"
END_OBJECT
OBJECT  = IMAGE_MAP_PROJECTION_CATALOG
^DATA_SET_MAP_PROJECT_CATALOG = 'DSMAPF.LBL'
DATA_SET_ID= 'MGN-V-RDRS-5-MIDR-FULL-RES-V1.0'
IMAGE_ID   = 'F-MIDR.10N076;1'
MAP_PROJECTION_TYPE      = SINUSOIDAL
MAP_RESOLUTION = 1407.4 <PIXEL/DEG>
MAP_SCALE   = 75 <M/PIXEL>
MAXIMUM_LATITUDE   = 12.5452
MAXIMUM_LONGITUDE = 73.8290
MINIMUM_LATITUDE   = 11.8187
MINIMUM_LONGITUDE = 73.0787
X_AXIS_PROJECTION_OFFSET = 17665
Y_AXIS_PROJECTION_OFFSET = 4096
X_AXIS_FRAMELET_OFFSET   = 1
Y_AXIS_FRAMELET_OFFSET   = 1
A_AXIS_RADIUS  = 6051.92 <KM>
B_AXIS_RADIUS  = 6051.92 <KM>
C_AXIS_RADIUS  = 6051.92 <KM>
FIRST_STANDARD_PARALLEL  = 0.0000
SECOND_STANDARD_PARALLEL = 'N/A'
POSITIVE_LONGITUDE_DIRECTION   = EAST
CENTER_LATITUDE   = 0.0000
CENTER_LONGITUDE  = 76.0583
REFERENCE_LATITUDE      = 'N/A'
REFERENCE_LONGITUDE     = 'N/A'
X_AXIS_FIRST_PIXEL      = 1
Y_AXIS_FIRST_PIXEL      = 1
X_AXIS_LAST_PIXEL = 1024
Y_AXIS_LAST_PIXEL = 1024
MAP_PROJECTION_ROTATION  = 0.0000
END_OBJECT
END
```

We learn from this file that the header record is 2048 bytes in size (2 records, each record is 1024 bytes). Moreover, the image identification is F-MIDR.10N076;1, image number 1 in a 56 image sequence centered at 10 degrees *N* latitude, 76 degrees *E* longitude on Venus. The size of this image is 1024 rows, each with 1024 pixels. The spatial resolution of each pixel is 75 m.

If digital images are in raw format without header records, information should accompany them documenting the size of the image in rows and pixels per row, and numeric resolution of the pixels, 8-bit, 16-bit, or some other resolution.

8.6.1.1   File Format Conversion

Visual_Data's first image processing tool, *Acquire an image for analysis,* is a simple image display tool that is also intended for the conversion of raw image data to Windows BMP format. The foregoing example of a Magellan radar image of Venus is revisited to show how this tool is used for file format conversion.

- start *Visual_Data,* click Tools, move the cursor to *Digital Image Analysis,* and click *Acquire an image for analysis;*
- The following questions are responded to interactively:

  1. select the image file to be opened (CD, directory \images\Venus\, file: FF01.IMG;
  2. respond yes or no if this image data is raw format
     a. if the answer is NO, the program assumes the format is BMP; the image is displayed and no further interaction is necessary;
     b. if the answer is YES, then
  3. are the pixels 16-bit resolution (if NO, the program assumes 8-bit);
  4. enter the number of header bytes: **2048 in the case of this image;**
  5. enter the number of rows: **1024, this image;**
  6. enter the number of pixels per row: **1024 this image;**
  7. finally, specify the name of the new BMP file to be created from the raw image.

Upon completion of Step 2A (for BMP images), or Step 7 (for raw images), the digital image is displayed. A portion of this display is shown below for this particular image of Venus.

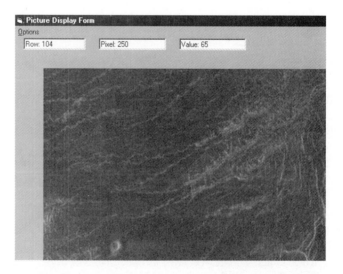

Notice that the upper part of this display shows row number, pixel number, and pixel value. Once the image is displayed, moving the mouse cursor onto the image changes the cursor icon to a cross (+). The row number, pixel number, and value is for the point in the image at the intersection of the cross. As the mouse is moved, new values appear in these boxes. This feature is necessary to digital image classification, a method of image visualization that is discussed in the last chapter of this text on data compositing.

8.6.1.2   Contrast Adjustment

Once the raw image is converted to BMP format, any one of the other image processing tools may be applied to its enhancement and analysis. In this example, contrast adjustment is used to enhance the display of the Venus radar image. From the digital image display form shown above, click Options and Return to the Main Program. Then, click Tools, move to Digital Image Analysis, and click Adjust Contrast. The following questions are responded to interactively:

1. the image file name is requested; this must be a BMP file;
2. a histogram of this image is displayed; the following is a histogram of the Venus image (this particular display is captured from the Statistical Analysis tool within the Digital Image Analysis group):

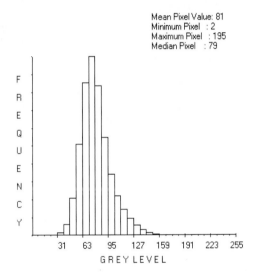

Notice that, whereas the Minimum pixel value is 2 and the Maximum pixel value is 195, the majority of pixel values are between 31 and 159. The histogram reveals a single mode (single distribution); multiple modes require more complex contrast transforms, such as multiple linear; single mode distributions are often more easily transformed using simple, linear stretches.

3. a user is then prompted to choose the type of enhancement that is desired:

   a. simple, linear: a user may choose the default; the program stretches the image between the minimum and maximum pixel values for the image that are shown at the top of the histogram display; or, a user may define the minimum and maximum values for the stretch. In the example of the Venus image, the following enhancement is chosen:

   ```
   Simple, linear; user defined; Min = 31; Max = 159.
   ```

   b. multiple, linear: a user may choose the default, a bilinear stretch: Min to Median and Median to Max, with Min and Max shown above the image histogram; or, a user may select as many linear segments as desired and defines the min and max values for each of these linear segments;

c. histogram equalization: this is an automatic conversion and no values are defined by the user;

d. custom: a user must first create a data file that contains the look-up table for the stretch; this is discussed earlier in this chapter, and shown explicitly in Section 8.6.3.when discussing Excel applications.

The outcome of the simple, linear stretch applied to the Venus image is shown below, along with the original image for comparison.

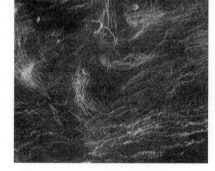

Original Image   Contrast Enhanced

8.6.1.3 Digital Filtering

Sigma (Lee) filtering is most commonly applied to microwave (radar) images in an effort to attenuate (suppress) speckle noise. This type of noise is unique to radar images acquired by synthesizing radar aperture. In this technique, a smaller antenna is used along with computer processing to sort Doppler frequency shifts caused by reflectance from objects whose positions relative to the antenna change as the antenna moves forward. But, many objects, not just a few, are imaged, so there are many reflected signals. These signals interfere with one another as they approach the antenna. Because these signals are electromagnetic waves, the interference is in the form of wave cancellations and additions. This causes darker (cancellation) and brighter (addition) pixels than what otherwise would occur for reflectance from a single object. This imparts a noise to radar images that is known as speckle.

Applying a sigma filter to a radar image involves experimentation to obtain a value, k, that appears to be appropriate. The size of the filter is fixed at 3×3. Several experiments are applied to the Venus radar image to show the influence of the parameter, k. *Visual_Data* is used to obtain these filtered results as follows:

1. From Tools, Digital Image Analysis is chosen, then Filter an Image is chosen, and Sigma (Lee) filtering is clicked.
2. After a brief delay as the program reads image data, a box prompts a user to specify the value of k; three experiments are attempted in this example for $k = 0.5$, 1.0, and 2.0;

3. A progress bar then tracks the progress of filtering. Digital filtering takes an amount of time that is proportional to the size of the image.

4. Once filtering is complete, a user is prompted to specify the filename for the new, filtered image; this file is created in Windows bitmap format.

Outcomes of the three experiments are shown below. Notice that image blur increases as k increases. A value, $k = 0.5$, diminishes speckle noise without causing severe image blur.

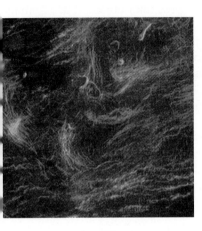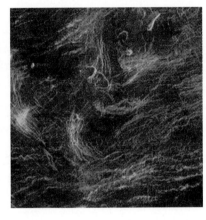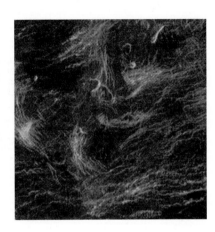

$k = 0.5$ $\qquad\qquad\qquad$ $k = 1$ $\qquad\qquad\qquad$ $k = 2$

Using *Visual_Data* for other types of filters is similar to what is described for sigma filtering. The type of filter is chosen and the user responds to questions interactively to obtain a filtered result. For example, suppose a high-pass filter is computed for this Venus image to highlight the structural features, folds and fractures, that are evident in this image. From the Filter group, high-pass filtering is clicked. In this type of filter, the size of the filter must be selected. Suppose a 5×5 filter is desired; the value, five, is simply entered to define the filter size because a square region of weights is assumed. The visual consequence is shown below:

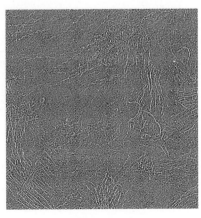

High-Pass Filter of
Unfiltered Image

High-Pass Filter, Sigma
Filtered, $k = 0.5$

Two high-pass filters are shown, not only to demonstrate this type of filter applied to radar images, but also to show the attenuation of speckle noise achieved through sigma filtering. High pass filtering emphasizes high frequency noises, such as speckle noise.

8.6.1.4 Image Addition and Subtraction

There are several ways to obtain filters of digital images. A direct way is through spatial convolution. An indirect way is through image subtraction. For example, a high-pass filtered image can be obtained as an image difference: original image − low-pass. If a low-pass filtered image is obtained using a 5 × 5 spatial convolution, the high-pass filter obtained using image differencing is optically equivalent to a 5 × 5 high-pass convolution. Another mathematical manipulation involves the recovery of the original image from high- and low-pass filterings: original image = high pass + low-pass. In this case, the high- and low-pass filters are obtained through convolution and the same filter sizes.

Using the Venus radar image, a 5 × 5 high pass filter is shown above. A 5 × 5 low-pass filter is computed using *Visual_Data*. Then, the Image differencing and addition tool is used to recreate the original image as follows:

1. Two image filenames are requested for the two images to be summed; or, if one image is to be subtracted from another in this order: C = A − B, the filename for image, A, is selected first, then that for image, B, is selected;
2. A delay occurs as the program accesses image data from both images; then, a user is prompted to select a difference or sum of the two images;
3. A user selects a file name for the new image (BMP format) as the final step. The experiment to recreate the original Venus image from high- and low-pass filters is shown below.

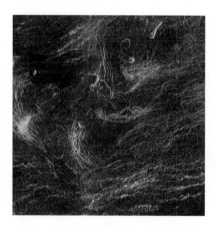
Low-Pass Filter (5 × 5)

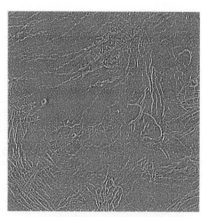
High-Pass Filter (5 × 5)

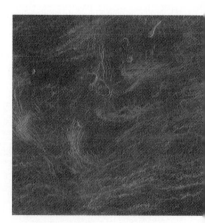
Sum: Low-Pass + High-Pass

8.6.2 ... Using MATLAB 5.3 (Student Version)?

MATLAB 5.3, Student Version, is shipped with all of the powerful tools that are shipped with the full, professional version. This includes a tool package for digital image processing. Up to now in this text MATLAB examples have been presented that have not depended on any of the supplied tools. In the present chapter, though, the digital image processing package is necessary and must be installed. When installing MATLAB 5.3, none of the additional tool packages is installed unless a user places a check in the box or boxes next to the desired package.

Consequently, to begin this section, if not installed already, the digital image processing tools must be installed. If using your own computer, one way to install additional tools is to first uninstall MATLAB, then reinstall the program, this time checking the additional tools that you wish to install. From Start, choose Programs, then MATLAB, then Uninstall. When finished, place the installation CD in the CD-ROM drive. The installation process should begin automatically. When the window is reached that shows the MATLAB components to install, scroll down the list until *Image Processing Tools* is reached. Place a check in this box and also a check in one of the help file boxes associated with it. The *HTML* help file version is convenient because it is accessible using any web browser program.

Once the installation process is completed, MATLAB is fully capable of all image processing functions described thus far in this chapter, and many additional applications. The simple display of an image is possible using the MATLAB function, *imshow.* The syntax is as follows (1 of several allowable syntax structures):

1. Start MATLAB
2. Change the directory to that containing a picture file in a compatible format (bmp, tif, gif, jpg).
   a. For example, to display one of the Mars Pathfinder images, place the CD supplied with this text in the CD-ROM drive. In MATLAB, type: cd [cd-rom drive letter]:\images\marspath\ and hit the enter key.
3. In MATLAB, type: imshow *filename*
   a. For example, type: imshow mistyblu.bmp
4. The following display is created:

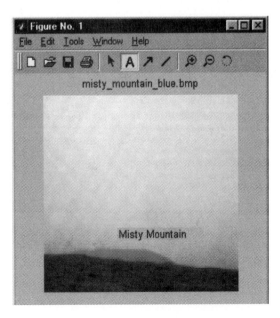

Notice the tools available with this image display. One is a magnifying glass with a + inside. This is an enlargement tool, enabling a user to zoom in on the image. Another tool, represented by the letter, A, allows text to be added to the image. The righthand display on the previous page demonstrates enlargement and text labeling. The magnifying glass, +, is clicked, the mouse cursor moved anywhere on the image and the left button is clicked. The letter, A, is clicked, the mouse cursor moved to where a label is desired, the left button clicked and the label is typed. Suppose that an image is not yet in one of the compatible formats such as bmp. In this case, a user of MATLAB may find it convenient to use Visual_Data's Acquire Image tool.

Other MATLAB image processing functions that match tools in *Visual_Data* are:

imhist   a MATLAB function that plots a histogram of an image; suggested syntax:

    a. imshow mistyblu

    b. I = getimage; (note: the semi-colon is part of the syntax)

    c. imhist (I, 32)

imadjust   adjust pixel values in an image (in other words, contrast enhancement); example syntax:

    a. I = imread('mistyblu.bmp');

    b. J = imadjust(I,[0.3 0.7],[]);

    c. imshow(I)

    d. figure, imshow(J)

Step d yields the following display:

conv2   a MATLAB function for two-dimensional convolution of matrices. The following example implements a low-pass filter:

    a. I = imread('mistyblu.bmp')

b. B = [0.11 0.11 0.11; 0.11 0.11 0.11; 0.11 0.11 0.11]

c. C = conv2(I, B);

d. J = uint8(C);

e. imshow (J)

The MATLAB function, uint8, converts C to single-byte (8 bit) integer data that is compatible with the MATLAB display function, imshow. Another example implements a high-pass filter, simply by changing step b to:

B = [−0.11 −0.11 −0.11; −0.11 0.88 −0.11; −0.11 −0.11 −0.11]

Steps a and c are as shown for the low-pass filter example. After step c, remaining

steps are:

d. C = C + 128; (Note: to shift negative values of C to positive)

e. J = uint8(C)

f. K = imadjust(J, [0.4 0.6], []); (Note: to enhance contrast)

g. imshow(K)

Step g yields the following display:

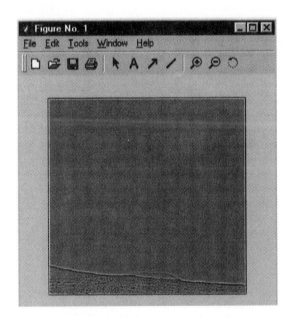

Other MATLAB image processing functions include: A. histeq, for histogram equalization contrast enhancement); B. medfilt2, for median filtering; and C. grayslice, for binary-type contrast stretching. As mentioned earlier, MATLAB offers many more image processing tools than does *Visual_Data*. Two functions in particular are demonstrated: fft2 and immovie.

Frequency domain filtering offers much more flexibility in removing image artifacts than does spatial convolution filtering. Earlier in this chapter in the discussion of Mars Pathfinder images, it is stated that Fourier domain filtering will not successfully remove the two-dimensional, cross-hatch noise associated with the CCD of the camera. A Fourier transform image of the blue filter image of Misty Mountain is used for demonstration. MATLAB computes a Fourier transform of this image as follows:

    a. I = imread('misty_blu.bmp');

    b. B = fftshift(fft2(I));

    c. imshow(log(abs(B)),[]), colormap(jet(64)), colorbar Step c yields the following, colorful display of the Fourier transform image: (See CD-ROM: \colfigs\ch8-fft.bmp)

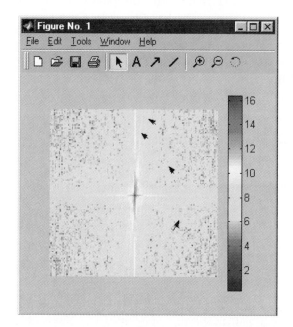

This display is enlarged and arrows added (using MATLAB's arrow tool with the display) to show several lines in the Fourier transform that are associated with the cross-hatch noise. If this noise was spatially regular, the same distance between lines, it would be representable by a single waveform. Consequently the Fourier transform would reveal it as a single line. In this case, many lines are associated with the cross-hatch noise, indicating its spatial irregularity. If regular, noise can easily be identified in the Fourier transform. This transform can be modified by zeroing out the part of it associated with the noise, and inverse transformation will yield the original image sans noise. When the noise is spatially irregular, however, isolating it in the Fourier transform is difficult or impossible. In this case, the noise is not easily removed. This is the case with the cross-hatch noise in the Mars Pathfinder images.

    Animation is an increasingly popular, and important form of data visualization. Animation is useful for simulating motion, or rapidly viewing data from many different perspectives. Once again using the Mars Pathfinder data as an example, MATLAB is used to create a movie of dust devil motion. As is discussed earlier in this chapter, the Mars Pathfinder camera imaged features at several electromagnetic frequencies, including visible frequencies, blue, green, and red. It did so by rotating filters. Consequently, some time elapsed between images obtained with the different filters. The elapsed time was approximately 20 seconds. The total time interval: blue imaging, green imaging red, imaging, spanned about 40 seconds. A dynamic feature, such as a dust devil, changes its position and geometry between filter rotations. Piecing these filters together in an animation enables the visualization of this motion.

The collection of images that will be used are from the IEEE Transactions on Geosciences and Remote Sensing article by Metzger, et. al. (2000). Three images make up this collection, sky-flat subtraction processed blue, green, and red filter images of Misty Mountain. MATLAB is used to create a move of these three frames through the following process:

a. I = imread('fig10b_blue.bmp');
b. J = imread('fig10b_green.bmp');
c. K = imread('fig10b_red.bmp');
d. X = moviein(3);
e. imshow(I);
f. X(1) = getframe;
g. imshow(J);
h. X(2) = getframe;
i. imshow(K);
j. X(3) = getframe;
k. movie(X, 10)

Step *k* plays the three frames in a movie 10 times. Change the value, 10, to any number of times for a longer or shorter movie. The three frames of the movie are shown below:

Blue Filter   Green Filter   Red Filter

8.6.3   . . . With the Help of *Microsoft Excel?*

Earlier in this chapter, the enhancement of digital image contrast is discussed. A demonstration showed how to implement a custom contrast stretch, in particular a custom binary stretch. Custom contrast enhancement using *Visual_Data* requires a user to first create a data file storing the 256 values of the look-up table that effects the custom stretch. This file is 256 lines long, with one number entered on each line. These numbers correspond to the following:

Line 1:   New output pixel value for input pixel value, 0
Line 2:   New output pixel value for input pixel value, 1
Line 3:   New output pixel value for input pixel value, 2
Line 256:   New output pixel value for input pixel value, 255

There are a number of ways in which a user may create this file. One way is to start *Visual_Data* and enter the 256 values in the upper box of the main window, saving the file using File, than Save (not Save As). This may be tedious and confusing, however, because *Visual_Data* does not number each line in the upper box; it is up to the user to keep count of lines. Consequently, a more convenient forum for file creation is a spreadsheet program that does show line numbers. One such program is *Microsoft Excel.*

In fact, *Excel* was used to create the look-up table that implemented the custom binary stretch that is shown earlier in this chapter. That stretch used a look-up table, designed as follows:

Lines 1 − 150 = 0
Lines 151 − 200 = 255
Lines 201 − 256 = 0

Microsoft Excel is used to create a data file for this look-up table using the following steps:

1. once *Excel* is started, the value, 0, is entered in Column A, lines 1 and 2;
2. using the mouse, the cursor is placed on the cell, Column A, line 1; the left mouse button is clicked and held down;
3. the mouse is then dragged down Column A to and including the cell for line 150; release the left mouse button;
4. click Edit on the main menu, then choose Fill, then click Down; all cells, Column A, lines 1 through 150 should be associated with the value, 0, if this step is completed successfully;
5. in Column A, lines 151 and 152, enter the value, 255;
6. moving the mouse cursor to the cell, Column A, line 151, click the left mouse button and hold it down, drag the mouse down Column A to and including the cell, line 200; release the left mouse button;
7. click Edit on the main menu, choose Fill, then click Down; all cells, Column A, line 151 through line 200, should be associated with the value, 255, if the step concludes successfully;
8. enter the value, 0, in Column A, lines 201 and 202; move the mouse cursor to the cell, Column A, line 201; hold the left mouse button down and drag the mouse to line 256; release the left mouse button; click Edit on the main menu, choose Fill, then click down; all cells, Column A, line 201 through line 256 should be associated with the value, 0.
9. click File, then choose Save As, and save this file as TAB DELIMITED TEXT. The look-up table is now complete.

8.7   Literature

An inspirational source for this chapter is Schowengerdt (1997), one of the first tomes on digital image processing that includes a discussion of geostatistics. Dust devils in Mars Pathfinder images were first reported in Metzger, et. al. (1998). Details on image processing of Mars Pathfinder images resulting in dust devil identification are presented in Metzger, et. al. (2000). The Venus radar image is taken from the CD-ROM, Magellan F-Mosaics MG_0066 V1, obtained from the National Aeronautics and Space Administration (NASA), Goddard Space Flight Center, National Space Data Center (NSDC).

Exercises

1. Write the look-up tables for the following stretches:
    a. Simple, linear; minimum = 35; maximum = 197.
    b. Bilinear: segment 1: min = 20, max = 72; segment 2: min = 72, max = 155.
    c. Binary; less than 130 is 0; greater than or equal to 130 is 255.

2. What is the optical result of the following look-up table?

Line 1:	255
Line 2:	254
Line 3:	253
:	
Line 256:	0

    Create this look-up table and apply it to an image, such as one of the *Nevada_Landsat* images and verify your conclusions.

3. What type of filter, high-pass or low-pass, is an Olympic filter? Using *Visual_Data*, apply this filter to one of the *Nevada_Landsat* images? Is the optical result consistent with your answer to this question? Explain.

4. One of the Viking images is shown below: Image ID: 038B25.

Notice the noise in this image, evident as random, black or white pixels scattered throughout this image. This is a spatially uncorrelated noise, and each aberrant pixel is isolated from others. This image is found on the CD-ROM at the end of this text in the directory, \images\viking\. Apply three filtering techniques to remove this noise: 3 × 3 low-pass; 3 × 3 high-pass; and 3 × 3 median. Which of these filters successfully removed this noise? Which of these filters enhanced this noise? Document what you learned from this experiment.

5. Adjust the contrast of your best filter result obtained in Problem 4. Carefully study this image. Can you identify dust devils?

6. The CD-ROM at the end of this text contains a directory, \images\marsglob\. Images and associated html files from this recent survey program of Mars are included. Digital image file formats are jpeg or gif, requiring conversion for analysis by *Visual_Data*. Use the Windows program, Paint, to convert the image, moc2_141a_msss, to Windows BMP file format. Paint displays this image. What features do you think are evident in this image?

7. Magellan radar images of Venus are written in the directory, \images\Venus\, on the CD-ROM. All are raw format. Choose any one of these images.
    a. Convert the image format to Windows BMP format.
    b. Adjust the contrast of the image
    c. Use sigma filtering to attenuate speckle noise.
    d. Analyze the image and identify all geologic features that are evident.

8. Three images of Twin Peaks are written to the directory, \images\marspath\, on your CD-ROM, one for each color filter, blue, green, and red. Using the analysis of Misty Mountain images as your guide, process the images of Twin Peaks. Did you discover a dust devil?

9. A ratio of two images is often used to enhance vegetation, for botanical studies, or iron staining when searching for mineral deposits. Using the suite of

Landsat TM images, *Nevada_Landsat_data*, prepare the following ratios:

a. Band 3 (numerator) over Band 1 (denominator)
b. Soil Adjusted Vegetation Index (SAVI); *Visual_Data* automatically computes this ratio as:

$$SAVI = (1 + L)(NIR - RED)/(NIR + RED + L)$$

L is a factor that ranges from 0 to 1. In humid environments associated with dense vegetation, L is 0, and SAVI is equivalent to NDVI, normalized difference vegetation index. In arid to semi-arid climes, L ranges from 0.5 to 1. Experiment with several values of L, 0; 0.5; 1.

10. Which of the ratios in Problem 9, A or B, is better when searching for iron staining? Why?

11. In problem 9B, which ratio enhances vegetation on alluvial fans better, one for which L is 0, L is 0.5, or L is 1?

Composite Visualizations

9

This final chapter of the text discusses data compositing to achieve unique visualizations. This type of data display is commonly referred to as raster-based (digital image based) *geographic information systems,* or GIS. Examples presented in this chapter do not represent a comprehensive summary of such visualizations. Instead, they represent a few examples to stimulate the understanding of data compositing. In general, the composite display of data is a visual approach to correlation analysis. Rather than computing a statistic, the correlation coefficient, color, perspective, and motion are used to infer correlation. Spatial aspects of this intercorrelation are also readily apparent by observing three-dimensional perspective. Through animation, we can achieve a better understanding of the dynamic characteristics of the processes that we are studying. Visualization through compositing offers an insight to data substantially different from that offered by a purely analytical treatment. Consequently, it is emerging as a method of increasing importance to the natural sciences.

9.1 Multispectral Digital Image Compositing for Classification

Composite display of digital images is introduced in Chapter 8. Image composites are used in the present chapter for multispectral classification. Another term for classification is **pattern recognition.** In this approach to data (image) visualization, features of interest are identified in a digital image. These features of interest are known as **classes.** A computer aided map is formed identifying all pixels in an image that belong to a particular class.

A class is distinguished by features known as **signatures.** For example, suppose we want to identify all pixels associated with vegetation. First, pixels known to be vegetation are identified. Computer software computes signatures for vegetation based on these pixels. One such signature is the histogram of pixels belonging to the class. Another is the mean. Yet one more is the variogram. All image pixels are compared to one or more of these signatures. Those that match closely are identified as vegetation and assigned a particular color. When finished, the result is a colorful map showing identified regions of vegetation. This is but one example. An analyst may select any feature of interest in an image for classification.

In multispectral classification, several spectral bands are composited enabling multispectral signatures to be identified. This improves the ability to distinguish different classes. For instance, water is associated with lower pixel values across all spectral bands. Vegetation, in contrast, is associated with lower pixel values in the visible portion of the electromagnetic spectrum, but significantly higher pixel values in the near infrared portion. Bare soil is associated with fairly large pixel values across the visible and infrared portions of the spectrum. These multispectral signatures are quite distinctive. By comparing the spectral characteristics of a pixel across bands to these signatures, its membership to one of these classes can readily be determined.

We learned in Chapter 4 that the *Nevada_Landsat_data* are associated with three prominent characteristics, water, vegetation, and native ground/sparse vegetation. Principal components transformation of these digital images separated these features well. Consequently, two multispectral classification experiments are attempted to classify these three image features. One experiment is based on a composite of bands 2, 3, and 4 (visible green, red, and near infrared), a **false color composite.** The other experiment is based on a composite of principal components images 1, 2, and 3 from both standardized principal components analysis and correspondence analysis. Results are compared to determine which composite is the more successful at distinguishing these three classes. Composites are displayed in Figure 9.1.

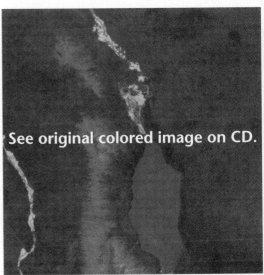

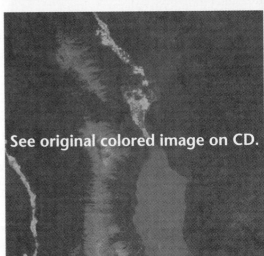

▲ FIGURE 9.1
Composite images. Clockwise from upper left: raw images, Landsat TM bands 2, 3, and 4; standardized principal components; and correspondence analysis.

Visual_Data implements classification using a **supervised** algorithm. In this technique, an analyst chooses **training sites** to define classes. Training sites are groups of pixels that an analyst knows to represent particular features, such as soil, water, or vegetation. It is up to the analyst to study an image to locate these groups of pixels. If using *Visual_Data*, the Acquire Image tool is useful to this process. Once the image is displayed, row number, pixel number, and pixel value are shown above the image. Dragging the mouse cursor across the image causes these values to change as the current mouse cursor position is changed. In *Visual_Data*, training sites are defined by the upper, left corner's row and pixel number, and the size (assumed square, an arbitrary programming choice) of the training site in pixels, for example 5 × 5 (25 total pixels), 10 × 10, and so on. The same training sites are used in these two classification experiments. They are identified in the following table.

	Training Site	**Test Site**	**Size (Training & Test)**
Vegetation	Row: 86 Pixel: 584	Row: 276 Pixel: 716	5 × 5
Water	Row: 886 Pixel: 893	Row: 1185 Pixel: 962	5 × 5
Native/sparse veg	Row: 629 Pixel: 1063	Row: 836 Pixel: 1065	5 × 5

Note: Row and Pixel numbers represent the upper, left corner location of training and test sites.

A minimum-distance-to-mean classification scheme is employed in these experiments. The mean (average) of the pixels in each training site for each spectral band is the signature on which these classification experiments depend. Once this process is completed for all training sites, classification proceeds by computing a "distance" for each pixel in the entire image. This metric is simply:

$$Dist_j = \sum_{i=1}^{N} |Pixel - Mean_{ij}|, \quad j = 1, 2, \ldots, M$$

in which N is the number of spectral bands, M is the number of different classes to be identified in the image, and Pixel is the numeric value of a pixel in an image. In multispectral classification, M total distance values are computed for each pixel. A pixel is ultimately assigned to the class for which $Dist_j$ is a minimum, hence the name of this particular algorithm.

This statement is tempered by the introduction of the concept of **thresholding.** What if $Dist_j$, although a minimum, is very large (a subjective assessment)? That is, what if this minimum distance is unacceptably large? Even though it is a minimum, the pixel may clearly not belong to the class. Thresholding is set by an analyst as that distance defining "unacceptably large." If the minimum distance is larger than the threshold, the pixel is not classified to any class and is displayed in black. In multispectral classification, the threshold is specified as a cumulative value over all spectral bands. For example, suppose that a 20-pixel distance in each spectral band is unacceptably large for the "minimum" distance, moreover suppose that three spectral bands are used for classification, then the threshold may be set at 60, the 20-pixel distance per band times the three bands. Thresholding is inherently experimental and this is simply one example. A small threshold value may severely restrict the classification and many pixels may go unclassified. A larger threshold value may result in all pixels being assigned to classes, even if they really don't belong to the classes to which they are assigned. In the experiments applied to the *Nevada_Landsat_data*, a threshold value of 60 is used (a pixel distance of 20 per band times the 3 bands). Figure 9.2 presents the outcome of these experiments.

Other than a visual inspection, there are several other ways to test the accuracy of classification. One technique relies on the use of **test sites,** regions of pixels in an image that an analyst suspects belong to the different classes. For unbiased testing, these test sites must not overlap training sites. The previous table listing training site locations also lists the locations of test sites. The objective is to determine how many of the test site pixels thought to belong to a class are actually assigned to that class. Results from this experiment are shown in the following tables:

Test site 1 (thought to be vegetation):
Results for both standardized principal components and correspondence analysis.

	Threshold	Vegetation	Water	Native
Threshold	8%	0%	0%	0%
Vegetation	0%	92%	0%	0%
Water	0%	0%	0%	0%
Native	0%	0%	0%	0%

Test site 1 (thought to be vegetation):
Results for classification based on raw image data:

	Threshold	Vegetation	Water	Native
Threshold	0%	0%	0%	0%
Vegetation	0%	100%	0%	0%
Water	0%	0%	0%	0%
Native	0%	0%	0%	0%

Test site 2 (thought to be water):
Results from all three composites:

	Threshold	Vegetation	Water	Native
Threshold	0%	0%	0%	0%
Vegetation	0%	0%	0%	0%
Water	0%	0%	100%	0%
Native	0%	0%	0%	0%

Test site 3 (thought to be native ground/sparse vegetation):
Results from all three composites:

	Threshold	Vegetation	Water	Native
Threshold	0%	0%	0%	0%
Vegetation	0%	0%	0%	0%
Water	0%	0%	0%	0%
Native	0%	0%	0%	100%

In terms of test site accuracy, classification based on raw image data appears to be better, at least for vegetation, in comparison to that which is based on principal components transforms. However, when comparing classifications visually (Figure 9.2), those based on principal components analysis

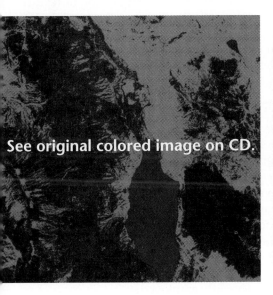

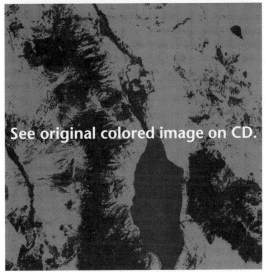

▲ FIGURE 9.2
Clockwise from upper, left: classification based on raw images; classification using standardized PC; classification using correspondence analysis. Note that the classification based on raw image data classifies pixels in shadow in the mountains to the west of Walker Lake as water. Principal components transformation enables a better distinction between water and shadow. The PC classifications differ from that based on raw data in the number of pixels that are assigned to native ground.

appear to be more accurate, at least with respect to the classification of native ground/sparse vegetation and water.

The best method of testing classifications is field checking. Sites are found on the ground that correspond to image pixel locations. Ground conditions are compared to pixel classification to determine accuracy. Of course, this requires actual travel to the field and access to locations identifiable in the classified image.

Foregoing examples illustrate classification based on spectral statistics. In particular, the spectral signature of a class in each band is its mean. Another classification algorithm, **Bayesian maximum likelihood,** uses the histogram of pixels in each band as the class signature for that band. Minimum-distance to-mean and Bayesian maximum likelihood algorithms are appropriately applied when the objective of classification is the identification of pixels of similar spectral characteristics.

9.1.1   Classification of Texture

Another approach to image classification is the identification of **texture.** In this case, the objective is to identify pixels that have similar spatial relationships. One type of relationship is spatial autocovariance. Consequently, the variogram is a useful signature. A variogram is computed for each textural class. If texture is isotropic, then an omnidirectional variogram is used. Otherwise, a directional variogram is useful for identifying texture that varies with direction.

Texture is a concept that is indeed difficult to visualize. In this demonstration, the "smooth" texture of water and the "banded, or striped" texture of ancient Walker Lake shorelines are classified. All pixels surrounded by pixels having a variogram that is similar to that for water will be classified as "water", or "smooth" texture. Likewise, those pixels surrounded by pixels having a variogram similar to the "banded, or striped shoreline" texture will be classified as shoreline. Because the shoreline texture is banded, exhibiting a pronounced directional tendency, an east-west calculation scheme is experimented with for the variogram. No directional tendency is observable for water (except some sediment patterns are noticeable, but are not so dominant as to warrant a directional variogram calculation), and an omnidirectional calculation scheme is consequently employed.

When classifying texture, the computer algorithm is considerably slower in comparison to spectral classification. A smaller subsection of the total image is extracted for this demonstration to reduce classification time. Training site locations for texture are shown in the following table; no test sites are chosen for this experiment.

	Row	Pixel	Size
Water	156	147	20 × 20
Shorelines	88	192	20 × 20

Note: Row and Pixel numbers represent the location of the upper, left corner of the training site. These row and pixel number locations are relative to the following, subsampled image, not the original full image used earlier. The subsampled image is displayed next to the classification map showing the result of textural classification.

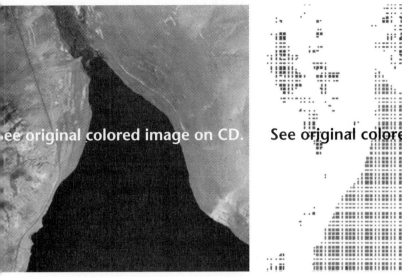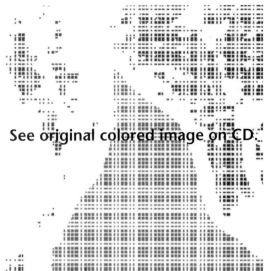

▲ FIGURE 9.3
Left image is an extraction from the full image. Right image is the classification map.
The texture of water is correctly classified; shorelines are as well, but tend to be con-
fused with alluvial surfaces.

Classifying texture or spectral characteristics depends on the objective of
the analysis. If the objective is the identification of mineralization (or rock
color in general), or vegetation health, or tonal variation, then spectral classification should be used. If, on the other hand, the objective is the identification of like spatial autocovariance, such as "smoothness/roughness" or
"striped ground" or "bedding", then textural classification is more appropriately applied.

9.2   Compositing a Contour Map
with a Digital Image

Reading Chapter 5, digital images are used to visualize results from kriging.
Another way to visualize these results is by drawing a *contour map*. *Visual_
Data* creates a text file of grid results, even if an analyst additionally chooses
to create a bitmap image of the grid. The grid text file is compatible with *Vi-
sual_Data's* contour line drawing tool once a new line is added as line 1 to define the following variables:

Nrow, Ncol, Nvalues, Ivalues, Iy, Ix

for which Nrow is the total number of rows in the grid, *Ncol* is the total number of columns in the grid, *Nvalues* is the total number of values on each line
in the grid text file (equal to 8 for a kriging outcome), *Ivalue* is that value on
each line that is to be contoured, *Iy* is that value on each line that defines the
y-coordinate, and *Ix* is that value on each line that defines the *x*-coordinate.

For example, recalling the *Nevada_Landsat_6x* examples used in Chapter
5, this first line of the grid file is 100, 100, 8, 5, 3, 4. This line is added to the
start of the grid text file as follows:

a. Once *Visual_Data* is started, from File, choose **open,** find the grid text
file and double click it;

 b. Make sure that the cursor is at the top, left of the file.

 c. Replace the line of text information at the beginning of the file with the one line shown on the previous page;

 d. From File, choose Save (not Save As), and save this file under its original name;

 e. From Tools, choose the contouring tool. The following display shows the contour map for the *Nevada_Landsat_6x* data set, visible blue.

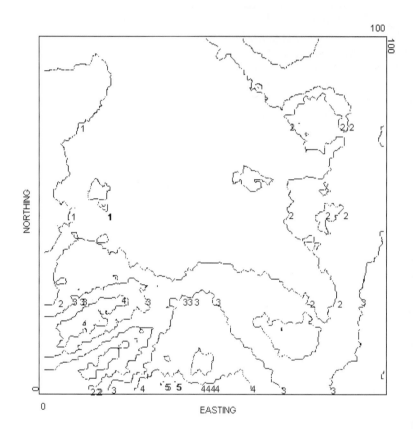

An experiment is attempted to superimpose this contour map onto the digital image for visible blue, *Nevada_Landsat_6x*. A digital image of the contour map is created once the contour map is displayed in *Visual_Data* by holding down the Alt key and hitting Print Screen. This pastes the graphic to the Windows clipboard. Then, from Start, choose Programs, then Accessories, then double click Paint. Once in Paint, click Edit, then Paste. Once the image is displayed in Paint, choose File, then Save As and save the image as a 256 color bitmap. The foregoing contour map display is actually a digital image created precisely as just described.

In order to display this contour map directly on the digital image, both bitmaps must be exactly the same size. Only by pure coincidence will the bitmap created by Paint match the digital image size. The simplest remedy to resize the contour map. The borders of the contour map should coincide with the margins of the digital image, consequently the only portion of the contour map image that should be saved is inside the borders. This image can be opened in Paint and the region inside the border clipped (cut). Using Paint's rectangle outline tool (its tool bar is shown on the left; the rectangle

ol is the upper, right of this toolbar), outlining that portion inside the bor-
r, clicking Edit and Cut, choosing File, then New (answer No to saving the
iginal image), then choosing Edit, then Paste displays the extracted por-
n of the contour map. Colors must be inverted for this compositing exper-
ent. White must become black, and black must become white. Inverting
lors in Paint is achieved by clicking Image on the main menu, then choos-
g Invert Colors. This result is then saved as a new 256 color bitmap image.
fore leaving Paint, Paint's pencil tool is clicked, moved to the lower, right
rner of the image and the row and pixel numbers are noted (these two
mbers are shown in the lower, right portion of the Paint window as the
ncil tool is moved around the image, but are respectively: pixel number
st, row number second).

Resizing of the contour map may be necessary. For instance, suppose the
gital image to be displayed with the contour map is 100×100. Further,
ppose the digital image of the contour map is 400×400. The contour map
ust be considerably downsized. Alternatively, the digital image is upsized.
n analyst chooses which of these two alternatives is desirable. *Visual_Data*
s a digital image resizing tool that can upsize or downsize a digital image
any desirable size.

In this example, two resizings are actually used. The digital image of vis-
le blue, *Nevada_Landsat_6x,* is upsized to 425×425. The contour map
own previously has an original size of close to 425×425. It is resized to
actly 425×425 to avoid an error when compositing with the digital image.
Note: the most common error if the two images are not precisely the same
ze is **Subscript Out of Range**). *Visual_Data's* digital image composite tool is
ed to create a 24 bit bitmap based on only two input images. The analyst
n choose which color will not be used when compositing only two images.
this case, the color blue, is not used. The contour map is loaded as green,
d the digital image is loaded as red. The following display shows the
ique visualization that is achieved when a line drawing is superimposed
a digital image. The display of line (vector) information on a digital image
commonly used for geographic information systems (GIS) applications.

Compositing a contour map with a digital image of data is useful for
veral reasons. This display allows the quality of the contour map to be
ecked. How well the contour lines represent actual changes in the data is
sily inspected. Moreover, higher and lower data values are more evident,
abling a better understanding of the contour map. Superimposing contour
es on the digital image is an alternative to a color bar, or legend for ex-
aining the numerical significance of pixel value.

.3   Three-Dimensional Perspectives

powerful form of visualization is the three-dimensional perspective of
idded information. A contour map accurately records data value, yet
ust be carefully studied to recognize data highs and lows. A digital image,
color or brightness, readily shows data highs and lows, but how rapidly
e data transition between these extremes is a bit difficult to visualize. Dis-
aying gridded information as three-dimensional relief reveals data highs
peaks, data lows as valleys, and the transition between as slopes of dif-
ring steepness.

An increasingly popular application of three-dimensional visualization is
e composite display of satellite data with digital elevation model (DEM)
ta. In this demonstration, a NASA/USGS cooperative product called

▲ FIGURE 9.4
Composite image. Contour map on digital image.

Pathfinder (not be confused with Mars Pathfinder) is used. One of the *Pathfin~~der~~* products includes *Walker Lake,* Nevada (Landsat TM Path 42, Row 33). Th~~is~~ product represents three time periods, 1970s, 1980s, and 1990s. Landsat da~~ta~~ from these time periods are registered to DEM data. The 1980 image is chose~~n~~ for demonstration. This image was acquired four (4) years prior to the Walk~~er~~ Lake images used throughout this text *(Nevada_ Landsat_data).*

Whereas Landsat data included in the *Pathfinder* product are written ~~in~~ single-byte (8 bit) pixels, the DEM data are written in higher radiometric pr~~e~~cision using two bytes per pixel (16 bits). Computer analysts using Window~~s~~ based personal computers should be aware that two-byte integers are re~~ad~~ differently than by other computer platforms. The byte order is reversed. ~~In~~ other words, if a 16-bit integer value is 0111000011111111, a Windows-base~~d~~ PC reads this as 1111111101110000. Consequently, it is necessary when using ~~a~~ Windows-based PC to swap byte order for some 16-bit images. (Note: this ~~is~~ not necessary for Mars Pathfinder 16-bit images, but is necessary for DE~~M~~ data). *Visual_Data* provides a tool for swapping byte order for 16-bit image~~s.~~

Once byte order is swapped, the following 8-bit image of the DEM da~~ta~~ for *Walker Lake* is obtained on the next page.

Digital elevation model data may be used alone when achieving uniq~~ue~~ three-dimensional visualizations. Most commonly, shaded relief images m~~ay~~ be created by using a variety of artificial sun positions and elevations. Visua~~l~~ Data provides a tool for three-dimensional grid display that allows artifici~~al~~ sun shading. The following image illustrates the creation of a shaded reli~~ef~~ image from the Walker Lake DEM image.

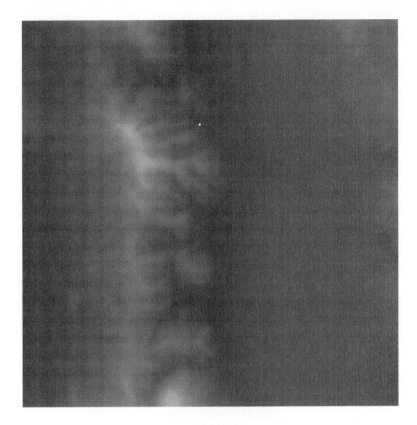

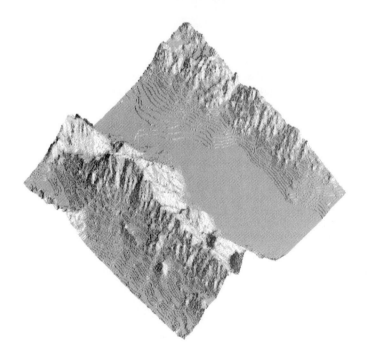

▲ FIGURE 9.5

Shaded relief image of *Walker Lake,* Nevada. This image was created using
Visual_Data, loading the elevation data as a bitmap image (the Walker Lake DEM
image). A tip angle of −45 and orientation of 225 yielded the orientation. A sun
angle position of 135 and sun elevation angle of −0.01 yielded the shading.

An alternative to using artificial sun position to compute shading inten sity, black to white, is to use Landsat pixel (color or intensity) to represent th shading. This is tantamount to draping the Landsat image onto the elevatic model. *Visual_Data* allows this type of "shading" to be used as an alternativ to sun angle shading. Suppose, for instance, Landsat TM band 2 is drape onto the DEM model. The following display illustrates the effect.

▲ FIGURE 9.6

Composite display of Landsat TM band 2 (visible green) and digital elevation, *Walker Lake,* Nevada. The DEM data provide the perspective of relief. The Landsat TM data define tone.

This type of composite display can be taken one step further by using 24-bit bitmap image to yield a display in which color and intensity of cole represent shading. Earlier in this chapter, composite displays based on prin cipal components transformation are presented. Vigorous vegetation assoc ated with agriculture appear brightly in these transformations. Is there a indication of natural vegetation activity in these images? The following is composite of digital elevation model with transformed images from corr spondence analysis.

Elevation data are a natural choice when creating three-dimensional di plays. But, unique three-dimensional visualizations are possible using othe types of data to define "elevation." When using kriging, for instance, two e timated quantities are obtained at each estimation location, an estimate value and kriging variance, an estimate of error. Suppose a digital image kriging variance is used to define elevation. In such a display, peaks repr sent higher kriging variance (error) and valleys represent lower kriging var ance. If a bitmap image of kriged estimates is draped on such an elevatic model, an analyst can recognize where the estimate bitmap is more suspe by noting the correlation with elevation peaks.

▲ FIGURE 9.7
Composite display of correspondence analysis color composite image with DEM
data. Vegetation at the north end of Walker Lake (north is up) has a purple-blue
color. The top of the mountain range west (to the left) of Walker Lake has a similar
color. In the Great Basin of North America, higher elevation is associated with
greater precipitation (in excess of 60 cm), consequently vegetation is more dense.
The dominant botanical species are conifers, having smaller leaf area and lower
near infrared reflectance in comparison to broader leafed agricultural species. This
type of composite display aids the interpretation of the digital image by understanding the spatial aspects of vegetation, natural and agricultural. Rock units to
the east of the lake are associated with unique color variation and consequently are
readily apparent and differentiable. This suggests an application of three-dimensional
visualization to geological interpretation.

A demonstration is developed using data, sanfer.dat (CD-ROM, directory
data\othrdata\). These are modified Mercalli intensity data for the 1971
San Fernando, California earthquake (Richter magnitude approximately
6.5). Separate images of kriged estimated value and kriging variance are presented. From these, two three-dimensional composites are formed, one that
uses kriging variance to define elevation, and the other in opposition that
uses kriged estimated value to define elevation.

Attention is called to the digital image of kriging variance. Where dark,
kriging variance is minimal, possibly indicating higher confidence in estimated value. In actuality, kriging variance is lower when sampling density is
higher. When an earthquake occurs in the United States (as well as many
other locations worldwide), questionnaires are distributed to residents within the geographic region affected by the seism. Sampling density is consequently greater within this geographic region than outside of it. The epicenter
of the 1971 San Fernando earthquake is within the darker region of the digital image of kriging variance. This image is an interesting one for observing
sampling pattern. Such an image may be used when deciding if, moreover
where additional sampling may be warranted. For dynamic processes, such

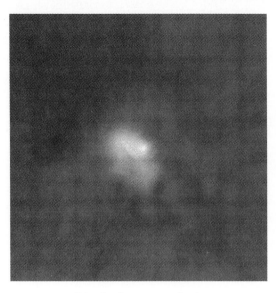

Kriged estimates, grey-level

Kriging variance, grey-level

as earthquakes, additional sampling likely is not possible, but for stat processes, additional sampling may be possible.

Three-dimensional composite images are formed using these two digit images. In the first composite, the digital image of kriging variance is used define elevation. The digital image of kriged estimate is used to defir "shading." The outcome is shown below.

This is an odd result. Yet, when observing the digital image of krigir variance, note that it is lower in the middle than at its margins. When used define elevation, the result is a depression in the center, like that of a funne Interesting perhaps, yet this display is difficult for use when inferring tl quality of estimated value.

An alternative display is pursued, this time using kriged estimate to d fine elevation, and kriging variance to define "shading." The outcome shown on the next page.

In this outcome, higher elevation is noted at the center of the image. Because elevation represents estimated ground motion intensity, higher
ground motion is evident in the center of this image. It has a dark shading.
Because kriging variance is used to define shading, we observe that the highest estimated value is associated with lower kriging variance. We may infer
higher confidence in these estimated values. This alternate display is better
used when examining the precision of estimated values.

These two experiments show how experimentation can lead to many,
different visualizations. A conceived visualization in our minds may not
work on paper. A good example is the first experiment, using kriging variance to define elevation. The intent was to show the confidence level in
kriged estimate. The funnel shaped visualization, however, is not convincing. What is more convincing is the visualization achieved in the second experiment. This underscores the fact that visualization is often a highly
empirical process.

9.4   Animation

Animation is introduced in Chapter 8 when discussing the use of MATLAB for
digital image processing. So impressive is MATLAB's capabilities in this regard that it motivates the modification of *Visual_Data* to provide an animation tool. Animation is used in Chapter 8 to visualize the motion of dust
devils across the Martian surface. Animation is now used to demonstrate the
usefulness of motion for recognizing certain features in data that may not
otherwise be obvious.

Animation is a unique type of data composite. A movie is created by
compositing many still photographs, each showing a slightly different perspective or position. When shown in rapid sequence, these still photographs
reproduce motion. This is animation.

In the present demonstration, the *Walker Lake* Pathfinder three-
dimensional composite is revisited. An animation is created to simulate fly-
ing over this lake, approaching from the east at a low elevation, ascending
to higher elevation, then moving toward the north. The focus is on the
mountain range to the west of Walker Lake. This mountain range is a horst

that is actively uplifting. Many triangular facets are noticed bordering Walke Lake. The animation sequence is used to highlight and study these triangula facets.

A sequence of twenty bitmap images are used, created by *Visua Data's* Three-Dimensional Gridding tool. The first 10 frames represent ti angles from −5 degrees up to −50 degrees, each with a spin angle of 0 de grees. Frames 11 through 15 have tip angles of −50, with spin angles ran ing from +5 degrees up to +20 degrees. Frames 16 through 20 have ti angles starting at −45 degrees descending to −25 degrees, with spin angle starting at +25 degrees up to +50 degrees. These variations are used t simulate flight.

Once these 20 frames are created, *Visual_Data's* animation tool is used create the movie. The total number of frames is specified and file names a specified one at a time. Finally, the analyst chooses how many times to pla the movie sequence. Animation cannot be duplicated on the pages of a te book, but the 20 frames are shown in the following figure.

When using animation for analysis, the challenges of movie directing a encountered. Frames are conceived and created as an analyst expects the to visually serve the animated sequence. Sometimes the visual effect is n quite what is intended, and the sequence is modified. In this demonstratio only twenty frames are used. In total, thirty minutes time was necessary fc their creation. By repeating this demonstration, better appreciation is rea ized for the amount of time and effort necessary when completing an enti movie, such as *Toy Story* and *Toy Story 2*, commercial movies based entirel on computer generated images.

9.5   How Do I Reproduce Results in This Chapter, or Pursue My Own Ideas . . .

9.5.1   . . . Using *Visual_Data?*

This chapter is written somewhat differently than previous chapters. Consic erable information is presented in this chapter explaining how to use *Visua Data* for creating the different visualizations. Still, more information can b presented showing how to reproduce results.

When compositing images into a 24 bit-bitmap image, *Visual_Data* give the user a choice of compositing two or three images. If compositing onl two images, such as a contour map and digital image, the user has the optic of which color to turn off (to not use). If, for example, the color, blue, is n used, the user simply enters a B (or b, lower case will work also) whe prompted for which color not to use.

If contouring data, *Visual_Data's* contour line drawing tool expects file that, as a minimum, contains information defining *x* and *y* coordina directions and the value to be contoured. The program is quite flexib and can accommodate many different files, provided they are in text (c ASCII) format. One line must be added at the beginning of the file th defines:

      Nrows   *Ncolumns*   *Nvalues*   *Ivalue*   *Iy*   *Ix*

These parameters are described earlier in this chapter. The following show the first five lines of a grid file that was modified for contouring:

FIGURE 9.8
Animated sequence comprised of 20 individual bitmap images.
Top to bottom, left to right; row 1; Frames 1–4 row 2; Frames 5–8 row 3; Frames 9–12
row 4; Frames 13–16 and row 5; Frames 17–20.

250	250	8	5	3		4		
1	1	249.5	0.5	5.00000007078052	13.23333	5	5.0000000596046
1	2	249.5	1.5	5.00000021606684	13.23333	5	5
1	3	249.5	2.5	5.00000021606684	13.23333	5	5.0000000596046
1	4	249.5	3.5	4.99999989569187	13.23333	5	5
1	5	249.5	4.5	4.99999989569187	13.23333	5	5

Note the first line added to this file. Lines two through five are the start of the grid file created by *Visual_Data's* kriging tool. There are eight values per line. The fifth value is the kriged estimate. The third value is the *y*-coordinate, and the fourth value is the *x*-coordinate. The total grid is 250 × 250. In summary, Nrows is 250, Ncolumns is 250, Nvalues is 8, Ivalue is 5, Iy is 3, and Ix is 4.

If using *Visual_Data's* digital image classification tool, an analyst must first select a training site, or sites (several may be used to define one class), for each and every desirable class. If testing is to be performed, test site locations must also be determined. The Acquire Image tool in *Visual_Data* is useful for studying an image to locate these sites. Once training (and testing) locations are known, the classification tool is started. Training (and testing) locations are specified interactively. The value of the threshold must also be specified. Choosing an optimal threshold may take some experimentation. If the initial classification result appears too conservative (few pixels are classified to any class), or too liberal (far too many pixels are classified), the threshold can be increased, for a conservative result, or decreased, for a liberal result, to adjust the classification. When experimenting with threshold, *Visual_Data* does not require information for training and test sites to be entered again.

For three-dimensional grid display, *Visual_Data* offers several options. A grid file from kriging can be entered directly, with a first line added that is identical to that required by the contour line drawing tool, except set Iy = row number position (1) and Ix = column number position (2). The grid can be displayed using lines, or artificial sun shading can be added. Tip and spin angles are used to orient the display. Tip should be a negative number to tip the display toward the computer analyst, a negative number from 0 (no tip) to −90. The tip angle giving a satisfying visual result depends on the data file being displayed and will likely require experimentation. The spin angle is specified as a compass direction, 0 to 360, or 0 to −360 degrees, and is set according to what aspect of the data an analyst wishes to emphasize.

If using sun shading, the sun position is specified, 0 to 360, or 0 to −360. Sun elevation angle must be a negative number to place the sun above the grid; smaller values, for example, −0.01, simulate very low sun angle. Sun position and elevation also require experimentation to achieve the desired optical effect. Suppose after specifying these values that the grid display is mostly dark. In this case, the sign of the sun position angle, negative or positive, should be reversed from that which was originally entered. The displays at the top of the next page show the *Nevada_Landsat_6x* grid, displayed with and without sun shading. Tip angle for each display is −45. Spin angle for each display is 225. Sun direction is 135. Sun elevation is −0.1.

If animation is the objective, bitmap images of each movie frame must be created first. A frame is simply one digital image. Once the total collection of bitmap images is assembled, *Visual_Data's* animation tool is started directly. All files are opened within this tool, one image at a time. The analyst then chooses how many times to repeat the movie sequence. The movie starts as soon as this prompt is responded to.

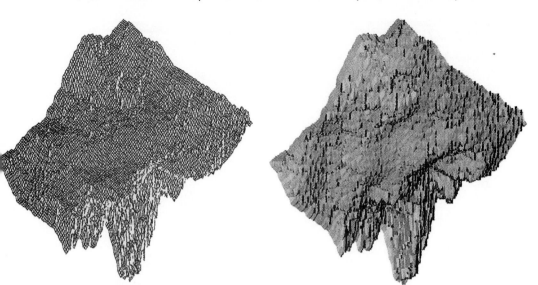

.5.2   . . . Using MATLAB?

ATLAB's digital image processing tools are introduced in Chapter 8. *Using* ATLAB for animation is described there. Its requirement for animation is the ame as for *Visual_Data*. Individual frames of the movie are first created as itmap images, then entered into the program. The syntax is as follows:

```
>>cd [directory containing movie frames]
>>for i = 1:Nframes
>>      imshow(X(i));        % X contains filenames
>>      M(i) = getframe;
>>end
>>movie(M, 30)               % plays movie, M, 30 times
```

MATLAB is also capable of contour mapping. Middleton (2000) presents ore sophisticated programs for this application. Herein, a simpler MATLAB rogram is presented, ctour.m, that translates a file from kriging (output om either *Visual_Data* or Pkrige_2.m) to one that is compatible with the ATLAB function, contour. This translation program is listed below:

```
function v = ctour(x, y, z, N, M)
% a function to convert kriging files to contour
% compatible files
for i = 1:N
 for j = 1:M
  k = (i - 1) * N + j;
  Z(i,j) = z(k);
  X(i,j) = x(k);
  Y(i,j) = y(k);
 end
end
contour(X,Y,Z), colorbar
```

Applying this translation program to the grid output from *Visual_Data* or the file, *Nevada_Landsat_6x*, results in the following contour display:

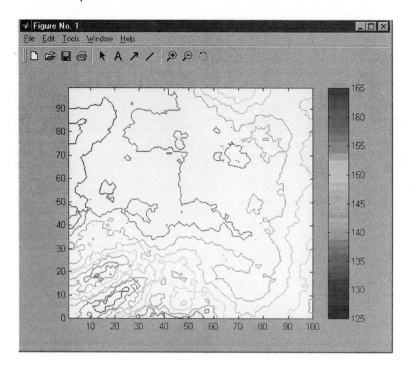

This contour map is created in MATLAB using the following steps:

```
>>cd [directory with grid file]
>>load filename          (example: load Nevada_Landsat_6x.grd)
>>W = filename           (example: W = Nevada_Landsat_6x)
>>x = W(:, ix)           (example: ix = 4 for Visual_Data or Pkrige_2.m grids
>>y = Nrow - W(:, iy)    (example: iy = 3 for Visual_Data or Pkrige_2.m grids
>>z = W(:, iz)           (example: iz = 5 for Visual_Data or Pkrige_2.m grids
>>cd [directory containing ctour.m] (example: cd q:\MATLAB)
>>ctour(x,y,z,N,M)       for a grid of N rows by M columns; e.g., N = M = 100
```

Another, simple MATLAB translation program is designed to render
gridding output from *Visual_Data* or Pkrige_2.m compatible with the MAT
LAB function, surf, a function that draws a three-dimensional surface wit
shading. This program is listed below:

```
function c = grid3d(x, y, z, N, M, Az, El, T, S)
% function to draw solid 3d elevation grid
% with sun shading, defined by Az and El
for i = 1:N
 for j = 1:M
  k = (i — 1) * N + j;
  Z(i,j) = z(k);
  X(i,j) = x(k);
  Y(i,j) = y(k);
 end
end
surf(X,Y,Z,'FaceColor','green','EdgeColor','none');
camlight(Az,El); lighting phong
view (-T, S)
```

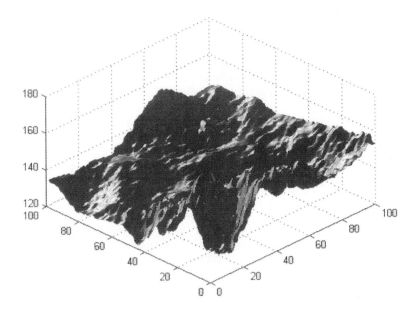

This program is used with exactly the same steps as shown above for ontouring, except the final step is:

```
>>grid3d (x,y,z,N,M,Az,El,T,S)
```

n which N is the row size of the grid, M is the column size of the grid, Az is he sun azimuth angle, 0 to 360, or 0 to −360, El is the sun elevation angle (en-ered as a *positive* number), T is the tip angle for the display, also entered as a *ositive* value, and spin is the spin angle, 0 to 360, or 0 to −360). For example, he following command is used to display the *Nevada_Landsat_6x* grid:

```
grid3d(x, y, z, 100, 100, −135., 20., 45., 45.)
```

his command yielded the display at the top of this page (green shading is sed).

Some final comments are forwarded about MATLAB for data compositing. MATLAB's capacity for compositing digital images into 24-bit bitmaps was not xplored. Its capability in this regard is not clear. Likewise unclear is MAT-AB's ability to drape a digital image onto data that define elevation. MATLAB lso does not appear to offer the ability to classify digital images.

.6   Literature

chowengerdt (1997) inspired classification experiments. *Visual_Data's* clas-ification routines are based on algorithms originally presented in Carr 1995). Three-dimensional visualization, particularly the draping of a digital nage onto elevation data, was inspired by a conversation with Eric Grun-ky, Alberta Geological Survey, Canada at the 1999 *American Geophysical Union* conference in San Francisco, California, on the significance of the fig-re that adorned the cover of the 1999 issues of *Computers and Geosciences*. hat figure shows a digital image of rock alteration draped on digital topo-raphic data for a region of Indonesia. Alteration is readily observed coin-iding with higher elevations, suggesting that erosion has removed altered ock at lower elevation. This aspect was not known prior to creating the hree-dimensional visualization. The three-dimensional grid models were eveloped based on algorithms originally presented by Angell (1985).

Exercises

1. The CD-ROM, directory\images, presents three directories for NASA Pathfinder data:
   a. Walker_ Lake;
   b. Lake_Tahoe;
   c. Pyramid_Lake_Nevada.

   Using the composite displays presented in this chapter, do one or more of the following exercises:
   a. Choose three different classes and classify a composite of bands 2, 3, and 4;use spectral classification and experiment with threshold; or, experiment with classification of texture;
   b. Repeat problem 1A, but use a composite of principal components images from standardized principal components, correspondence analysis, or both; what about mixing images from spca and correspondence analysis?
   c. Create three-dimensional grid models by draping one or more of these digital images onto the digital elevation model data. The DEM data must be converted to 8-bit bitmaps. They are presently written in 16-bit integer format, and their byte-order must be swapped.
   d. Create a movie of a three-dimensional grid model.

2. Another application of digital image compositing is **change detection.** Two digital images acquired at different times are geographically registered. Geographic registration is a process in which two images are available, and one is shifted such that its pixel positions coincide exactly with those of the other image such that they can be displayed coincidentally. The NASA Pathfinder images and DEM data are geographically registered. In each Pathfinder directory, an additional Landsat image from the 1990s is included. This image is a band four, infrared image. Experiment with as many different techniques as you can imagine for showing changes in these three geographic areas:

   Walker Lake, Nevada
   Lake Tahoe, Nevada/California
   Pyramid Lake, Nevada

3. Recalling Chapter 7, create some two-dimensional, nonconditional simulations. Save them as grey-level bitmap images. Use these images to define elevation in the three-dimensional gridding tool. Apply sun shading. Are these good models for topography? Explain.

4. Create an animation of the sun moving across a surface from dawn to dusk.

5. Apply kriging to one of the data sets in the directory, \data\othrdata\.
   a. Create a contour map and composite its image with a bitmap image of kriged estimates;
   b. Create a three-dimensional composite of kriged estimate and kriging variance. Which visualization do you prefer?

   Note: a readme file in the directory, \data\othrdata\, describes these data. Consult this file when choosing data set of interest for analysis.

Epilogue

Why Be Normal?

That a normal distribution model was inappropriate for any one of the seven spectral bands comprising the *Nevada_Landsat_data* did not preclude the application of any analytical method described in this text. Many of these methods are nonparametric, not dependent upon the presumption of normality. Many of these methods are robust, insensitive to data distribution. Robustness is often controlled by the data analyst. This is particularly true with kriging. Parameters chosen to effect estimation, especially grid geometry, variogram range, and number of nearest neighbors, impact robustness. Data analysts also influence robustness when using other methods, particularly regression analysis. A great deal of insight was rendered from the *Nevada_Landsat_data* by the various analytical methods described in this text, despite the nonnormality of these data. Too much emphasis is placed on the normal distribution model in analyses of natural data?

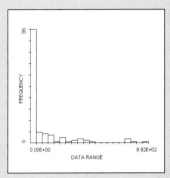

Human Prerogative

Mathematics paradoxically frightens and comforts. A student of geology, perhaps intimidated by mathematics, is excited by the pronounced **correlation coefficient** of 0.85 between lead and silver in a deposit. The mathematical premise of correlation may be daunting, but a *number* is obtained that is important to the student. The deposit contains galena, lead sulfide, that often has a high silver content. The relatively high value of correlation coefficient suggests that this is the case for this deposit. Although the math is frightening to the student, the number is comforting for its support of a scientific conclusion.

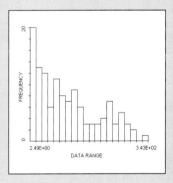

What about the circumstance in which the number does not support the inference? Is there an urge, in this case, to dismiss the mathematics and support the inference? This urge must be suppressed. The conflict must be resolved. For instance, the inference may be flawed, or the data on which correlation analysis is based may be in error.

Early in this text, analysis of correlation among the seven spectral bands of the *Nevada_Landsat* data yielded coefficients that suggested fair to good correlation. Scatter plots of some of these relationships, though, yielded strange patterns that seemed to contradict the correlation coefficients. Not much advice was presented on how to deal with this visual contradiction. Deference was given then to the mathematical outcome.

Is there a human prerogative to adjust the mathematical outcome upon visual inspection of the scatter plot? Recall, for instance, the scatter plot between visible blue and near infrared reflectance that was shown in Chapter 3.

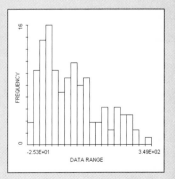

Mathematically, a correlation coefficient of 0.67 is obtained based on th "best-fit" line. Indeed, this value was used in Chapter 4 for principal compo nents analysis. A visual inspection of this scatter-plot could provoke the opir ion that this correlation coefficient is too high. In fact, there is some visua suggestion in the scatter-plot for *negative* correlation. Upon reflection, it is re membered that healthy, living vegetation has strong NIR reflectance and lov blue reflectance. Shallow water may have a high blue response, but very lov NIR response. On the basis of these relationships, a negative correlation coef ficient makes sense. There is justification, however, for positive correlatior Deep water has lower blue and NIR responses. Bare soil has higher blue an NIR responses. This enigma necessitates human intervention. There is human prerogative to overrule the mathematics.

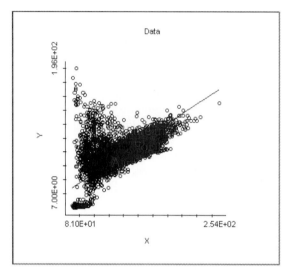
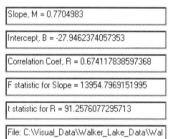

▲ FIGURE 10.1

Recall the analysis of correlation between visible blue (*x*-axis) and near infrared (*y*-axis) reflectance.

In principal components methods, though, a single number is used t represent the degree of correlation between two variables. This scalar quar tity, whether positive or negative, cannot simultaneously represent the d chotomous negative and positive relationships that seem to exist betwee visible blue and near infrared reflectance (as one example). Human interver tion in the principal components analysis process may provide useful insigh to the value of correlation metric yielding the most "useful" visual results.

For example, suppose the correlation coefficient matrix in standardize principal components analysis is modified to render the correlation coeff cient between visible blue and near infrared reflectance equal to −0.67. Th absolute correlation is that which was used before, but the sign is opposite In this case, the negative relationship between these two spectral frequencie is emphasized. The graphical result is displayed below, along with the orig nal result from Chapter 4 for comparison.

Relative variable relationships are unchanged. In general, emphasis o the negative relationship between visible blue and near infrared reflectanc strengthens the relationship between visible green and red reflectance. Sam ples (shown by number identification in Figure 10.2) are less distinguishabl

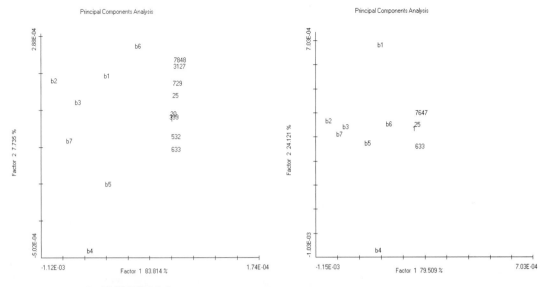

▲ FIGURE 10.2
Left figure shows standardized principal components analysis of *Nevada_Landsat_6x_data*, originally presented in Chapter 4. Right figure shows the analysis after changing the correlation coefficient between visible blue and near IR reflectance from 0.67 to −0.67.

f the goal is sample separation, the original, positive correlation between
blue and near IR reflectance is better used. If, on the other hand, a goal of this
analysis is better discrimination between visible blue and the other visible
reflectance, green and red, the negative correlation between blue and near IR
is better for meeting this objective. In general, when confronted by visually
complex, bivariate data relationships, experiments such as this one aid the
understanding of what aspect of the two-variable behavior is most important to the data analysis. Consequently, there is a human prerogative to over-
ride the mathematical assessment.

s Kriging the Best Spatial Interpolation Method?

This text has, on occasion, accidentally or intentionally revealed the limita-
tions of powerful, mathematical algorithms. A notable example is cokriging.
Although seemingly more sophisticated than kriging, cokriging is better applied than kriging only in the instance when higher resolution data are used
o estimate lower resolution data.

There are alternatives to kriging. Polygonal interpolation and inverse
distance weighting are two, widely used algorithms. Polygonal interpola-
ion sets the value of an estimate equal to the closest, neighboring data value.
This is tantamount to collapsing the kriging neighborhood to $N = 1$ closest
data locations. Inverse distance weighting applies weights to closest neighboring data locations solely on the basis of their physical distance from an
stimation location. The closer a data location is, the larger is the weight applied to it.

Kriging, inverse distance weighting, and polygonal estimation use the
ame equation when computing an estimate: $Z^*(x_0) = \sum \lambda_i Z(x_i)$. All three
re local estimators; that is, they use the closest data values when computing

estimates. They differ with respect to the determination of weights. Kriging computes weights as a function of spatial autocorrelation. Inverse distance weighting is purely geometric with weights computed as a function of the inverse of distance from an estimation location. Polygonal estimation involves $\lambda = 1$, applied to the one, closest location to an estimation location. How significant, really, are these differences?

A demonstration is presented, based on the analysis of data from Englund (1990). Kriging and inverse distance weighting are based on using the ten nearest neighboring values. Notice the similarities and differences among these methods shown below. (Data: area1.dat, \data\othrdata).

  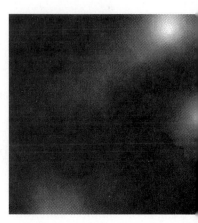

Polygonal   Inverse Distance   Kriging

▲ FIGURE 10.3

A comparison of three interpolation algorithms in application to the same data set.

Polygonal estimation more dramatically highlights regions of higher data value. Inverse distance weighting pinpoints higher data values rather precisely, at least in the case of this one data set. Kriging provides a smoother result. Combined, these three visualizations offer a more complete understanding of the spatial distribution of higher and lower data value than what is obtainable through the application of any one.

Color compositing and animation are further useful for visualizing the similarities and differences among these three algorithms. Animation, playing the three images shown in Figure 10.3 over and over again, aids the assessment of the spatial distribution of higher data value. Color compositing (Figure 10.4) shows that polygonal interpolation tends to over emphasize higher data value, whereas there is more parity between inverse distance weighting and kriging. When attempting a more complete spatial analysis because each of these three algorithms offers a unique, visual perspective, their combined use enables an assessment of smoothing and more precise visualization of higher data value. Which of these algorithms is deemed "best" depends on the data, objectives of the analysis, and the analyst's opinion.

The Robustness of Kriging

Final thoughts are presented regarding the robustness of kriging. Attention was focused in Chapter 5 on the influence of the number of nearest, neighboring data locations, N, used in kriging on robustness. Cross validation is applied to the data from Englund (1990), varying N from 10 to 120. Results are shown in Figure 10.5. (Data: area1.dat, \data\othrdata).

▲ FIGURE 10.4

Composite display created by forming a 24-bit bitmap, polygonal estimation as blue, inverse distance estimation as green, and results from kriging as red. Bluer cells around higher data values shows the tendency of polygonal interpolation for over-estimation.

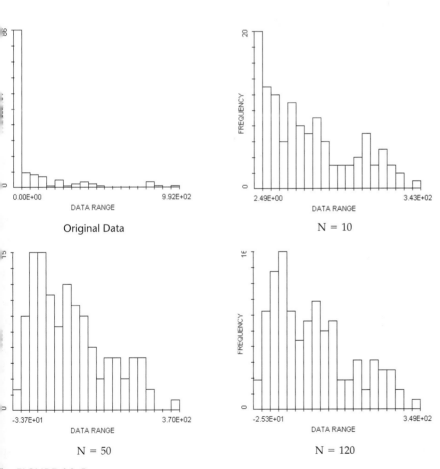

FIGURE 10.5

Histograms of estimated value after application of kriging. Original histogram shown for comparison.

Notice that as the kriging neighborhood is expanded (N is larger), the distribution of estimated value becomes more normal in appearance. The larger N is, the more necessary it is that the normal distribution model represents well the distribution of the raw data. This is tantamount to the claim that the robustness of kriging is lesser for larger N.

Appendices

APPENDIX A

CRITICAL VALUES OF THE CHI-SQUARE DISTRIBUTION

Table A-1	Chi-Square distribution for selected degrees of freedom and selected cumulative probability levels									
df	0.005	0.010	0.025	0.050	0.100	0.900	0.950	0.975	0.990	0.995
1	0.0000	0.0002	0.0010	0.0039	0.0158	2.7055	3.8415	5.0239	6.6349	7.8795
2	0.0100	0.0201	0.0506	0.1026	0.2107	4.6052	5.9915	7.3778	9.2103	10.5966
3	0.0717	0.1148	0.2158	0.3518	0.5844	6.2514	7.8147	9.3484	11.3449	12.8382
4	0.2070	0.2971	0.4844	0.7107	1.0636	7.7794	9.4877	11.1433	13.2767	14.8603
5	0.4117	0.5543	0.8312	1.1455	1.6103	9.2364	11.0705	12.8325	15.0863	16.7496
6	0.6757	0.8721	1.2373	1.6354	2.2041	10.6446	12.5916	14.4494	16.8119	18.5476
7	0.9893	1.2390	1.6899	2.1674	2.8331	12.0170	14.0671	16.0128	18.4753	20.2778
8	1.3444	1.6465	2.1797	2.7326	3.4895	13.3616	15.5073	17.5346	20.0902	21.9550
9	1.7349	2.0879	2.7004	3.3251	4.1682	14.6837	16.9190	19.0228	21.6660	23.5893
10	2.1559	2.5582	3.2470	3.9403	4.8652	15.9872	18.3070	20.4832	23.2093	25.1882
11	2.6032	3.0535	3.8157	4.5748	5.5778	17.2750	19.6751	21.9201	24.7250	26.7569
12	3.0738	3.5706	4.4038	5.2260	6.3038	18.5493	21.0261	23.3367	26.2170	28.2996
13	3.5650	4.1069	5.0088	5.8919	7.0415	19.8119	22.3620	24.7356	27.6882	29.8194
14	4.0747	4.6604	5.6287	6.5706	7.7895	21.0641	23.6848	26.1190	29.1413	31.3194
15	4.6009	5.2293	6.2621	7.2609	8.5468	22.3071	24.9958	27.4884	30.5779	32.8013
16	5.1422	5.8122	6.9077	7.9616	9.3122	23.5418	26.2963	28.8454	32.0001	34.2674
17	5.6972	6.4078	7.5642	8.6718	10.0852	24.7690	27.5871	30.1910	33.4085	35.7182
18	6.2648	7.0149	8.2307	9.3905	10.8649	25.9894	28.8693	31.5264	34.8053	37.1564
19	6.8440	7.6327	8.9065	10.1170	11.6509	27.2036	30.1435	32.8523	36.1907	38.5820
20	7.4338	8.2604	9.5908	10.8508	12.4426	28.4120	31.4104	34.1696	37.5662	39.9968
21	8.0337	8.8972	10.2829	11.5913	13.2396	29.6151	32.6706	35.4789	38.9322	41.4011
22	8.6427	9.5425	10.9823	12.3380	14.0415	30.8133	33.9245	36.7808	40.2895	42.7960
23	9.2604	10.1957	11.6886	13.0905	14.8480	32.0069	35.1724	38.0755	41.6382	44.1808
24	9.8862	10.8564	12.4012	13.8484	15.6587	33.1962	36.4150	39.3641	42.9798	45.5586
25	10.5197	11.5240	13.1197	14.6114	16.4734	34.3816	37.6525	40.6466	44.3144	46.9285
26	11.1602	12.1981	13.8439	15.3792	17.2919	35.5632	38.8852	41.9233	45.6419	48.2903
27	11.8076	12.8785	14.5734	16.1514	18.1139	36.7412	40.1133	43.1946	46.9631	49.6452
28	12.4613	13.5647	15.3079	16.9279	18.9392	37.9159	41.3372	44.4608	48.2783	50.9936
29	13.1211	14.2565	16.0471	17.7084	19.7677	39.0875	42.5570	45.7224	49.5881	52.3360
30	13.7867	14.9535	16.7908	18.4927	20.5992	40.2560	43.7730	46.9793	50.8922	53.6720
31	14.4578	15.6555	17.5387	19.2806	21.4336	41.4218	44.9854	48.2320	52.1915	55.0030
32	15.1340	16.3622	18.2908	20.0719	22.2706	42.5847	46.1942	49.4804	53.4856	56.3278

Table A-1 Chi-Square distribution for selected degrees of freedom and selected cumulative probability levels (continued)

df	0.005	0.010	0.025	0.050	0.100	0.900	0.950	0.975	0.990	0.995
33	15.8153	17.0735	19.0467	20.8665	23.1102	43.7451	47.3998	50.7250	54.7753	57.6479
34	16.5013	17.7891	19.8063	21.6643	23.9523	44.9031	48.6023	51.9659	56.0608	58.9636
35	17.1918	18.5089	20.5694	22.4650	24.7967	46.0588	49.8019	53.2035	57.3424	60.2755
36	17.8867	19.2327	21.3359	23.2686	25.6433	47.2122	50.9985	54.4373	58.6193	61.5813
37	18.5858	19.9602	22.1056	24.0749	26.4921	48.3635	52.1925	55.6683	59.8932	62.8847
38	19.2889	20.6914	22.8785	24.8839	27.3430	49.5127	53.3837	56.8958	61.1628	64.1827
39	19.9959	21.4262	23.6543	25.6954	28.1958	50.6598	54.5724	58.1203	62.4287	65.4767
40	20.7065	22.1643	24.4330	26.5093	29.0505	51.8051	55.7586	59.3420	63.6914	66.7673
41	21.4208	22.9056	25.2145	27.3256	29.9071	52.9485	56.9424	60.5606	64.9501	68.0528
42	22.1385	23.6501	25.9987	28.1441	30.7654	54.0902	58.1241	61.7769	66.2065	69.3365
43	22.3595	24.3976	26.7854	28.9647	31.6255	55.2301	59.3034	62.9902	67.4589	70.6151
44	23.1387	25.1480	27.5746	29.7875	32.4871	56.3685	60.4808	64.2014	68.7093	71.8921
45	24.3110	25.9013	28.3662	30.6123	33.3504	57.5053	61.6563	65.4102	69.9570	73.1663
46	25.0413	26.6572	29.1600	31.4390	34.2152	58.6405	62.8295	66.6162	71.2007	74.4353
47	25.7746	27.4158	29.9562	32.2676	35.0814	59.7742	64.0010	67.8204	72.4428	75.7031
48	26.5106	28.1770	30.7545	33.0981	35.9491	60.9065	65.1706	69.0223	73.6820	76.9675
49	27.2493	28.9406	31.5549	33.9303	36.8182	62.0375	66.3386	70.2224	74.9195	78.2307
50	27.9907	29.7067	32.3574	34.7643	37.6886	63.1671	67.5048	71.4201	76.1537	79.4896
51	28.7347	30.4750	33.1618	35.5999	38.5604	64.2954	68.6693	72.6160	77.3860	80.7468
52	29.4812	31.2457	33.9681	36.4371	39.4334	65.4224	69.8321	73.8098	78.6156	82.0005
53	30.2300	32.0185	34.7763	37.2759	40.3076	66.5482	70.9934	75.0018	79.8432	83.2522
54	30.9813	32.7935	35.5863	38.1162	41.1830	67.6729	72.1534	76.1924	81.0697	84.5036
55	31.7348	33.5705	36.3981	38.9580	42.0596	68.7962	73.3115	77.3805	82.2923	85.7493
56	32.4905	34.3495	37.2116	39.8013	42.9373	69.9187	74.4686	78.5677	83.5146	86.9960
57	33.2484	35.1305	38.0267	40.6459	43.8162	71.0397	75.6238	79.7523	84.7330	88.2368
58	34.0084	35.9134	38.8435	41.4919	44.6960	72.1596	76.7774	80.9349	85.9486	89.4738
59	34.7704	36.6983	39.6619	42.3393	45.5770	73.2791	77.9309	82.1181	87.1674	90.7185
60	35.5345	37.4849	40.4818	43.1880	46.4589	74.3972	79.0823	83.2984	88.3810	91.9547
61	36.3005	38.2732	41.3031	44.0379	47.3418	75.5139	80.2318	84.4758	89.5898	93.1833
62	37.0684	39.0633	42.1260	44.8890	48.2257	76.6300	81.3807	85.6531	90.8001	94.4159
63	37.8382	39.8551	42.9503	45.7414	49.1106	77.7456	82.5291	86.8302	92.0115	95.6520
64	38.6098	40.6486	43.7760	46.5949	49.9963	78.8597	83.6753	88.0042	93.2172	96.8787
65	39.3831	41.4436	44.6030	47.4496	50.8829	79.9728	84.8203	89.1764	94.4204	98.1019
66	40.1582	42.2402	45.4314	48.3054	51.7705	81.0856	85.9650	90.3492	95.6263	99.3315
67	40.9350	43.0384	46.2610	49.1623	52.6589	82.1974	87.1085	91.5202	96.8298	100.5579
68	41.7135	43.8380	47.0920	50.0202	53.5481	83.3078	88.2501	92.6884	98.0280	101.7751
69	42.4935	44.6392	47.9242	50.8792	54.4381	84.4180	89.3914	93.8568	99.2284	102.9978
70	43.2751	45.4417	48.7575	51.7393	55.3289	85.5269	90.5310	95.0227	100.4242	104.2130
71	44.0583	46.2456	49.5921	52.6003	56.2205	86.6346	91.6688	96.1860	101.6153	105.4204
72	44.8430	47.0510	50.4279	53.4623	57.1129	87.7426	92.8074	97.3515	102.8128	106.6410

Table A-1   Chi Square distribution for selected degrees of freedom and selected cumulative probability levels (continued)

df	0.005	0.010	0.025	0.050	0.100	0.900	0.950	0.975	0.990	0.995
73	45.6292	47.8577	51.2648	54.3253	58.0060	88.8492	93.9441	98.5140	104.0046	107.8518
74	46.4169	48.6656	52.1028	55.1891	58.8999	89.9550	95.0796	99.6749	105.1942	109.0595
75	47.2060	49.4750	52.9419	56.0540	59.7945	91.0614	96.2165	100.8391	106.3924	110.2845
76	47.9965	50.2856	53.7821	56.9198	60.6898	92.1661	97.3509	101.9992	107.5823	111.4950
77	48.7883	51.0974	54.6233	57.7864	61.5858	93.2692	98.4826	103.1548	108.7632	112.6892
78	49.5815	51.9104	55.4655	58.6539	62.4824	94.3725	99.6151	104.3126	109.9503	113.8963
79	50.3761	52.7246	56.3089	59.5222	63.3798	95.4757	100.7478	105.4712	111.1404	115.1097
80	51.1719	53.5401	57.1532	60.3914	64.2778	96.5782	101.8794	106.6284	112.3284	116.3203
81	51.9690	54.3566	57.9984	61.2614	65.1764	97.6790	103.0085	107.7816	113.5083	117.5165
82	52.7673	55.1743	58.8446	62.1322	66.0757	98.7798	104.1378	108.9355	114.6908	118.7184
83	53.5668	55.9930	59.6917	63.0038	66.9755	99.8793	105.2651	110.0864	115.8674	119.9100
84	54.3676	56.8130	60.5398	63.8763	67.8761	100.9799	106.3946	111.2418	117.0556	121.1244
85	55.1696	57.6339	61.3888	64.7494	68.7771	102.0789	107.5217	112.3933	118.2355	122.3242
86	55.9727	58.4559	62.2386	65.6233	69.6788	103.1772	108.6478	113.5435	119.4137	123.5213
87	56.7769	59.2789	63.0893	66.4978	70.5810	104.2744	109.7722	114.6908	120.5862	124.7085
88	57.5823	60.1030	63.9409	67.3732	71.4838	105.3722	110.8979	115.8413	121.7668	125.9119
89	58.3887	60.9280	64.7933	68.2493	72.3871	106.4687	112.0216	116.9883	122.9405	127.1031
90	59.1963	61.7540	65.6465	69.1259	73.2910	107.5642	113.1438	118.1331	124.1099	128.2867
91	60.0049	62.5810	66.5006	70.0034	74.1955	108.6601	114.2671	119.2804	125.2860	129.4841
92	60.8145	63.4090	67.3555	70.8816	75.1005	109.7555	115.3895	120.4266	126.4605	130.6790
93	61.6252	64.2378	68.2112	71.7603	76.0058	110.8490	116.5089	121.5675	127.6238	131.8534
94	62.4369	65.0676	69.0676	72.6397	76.9118	111.9430	117.6295	122.7111	128.7941	133.0417
95	63.2496	65.8983	69.9248	73.5198	77.8184	113.0373	118.7509	123.8567	129.9698	134.2410
96	64.0633	66.7299	70.7828	74.4005	78.7254	114.1304	119.8704	124.9991	131.1389	135.4288
97	64.8780	67.5624	71.6415	75.2818	79.6329	115.2231	120.9894	126.1411	132.3080	136.6170
98	65.6935	68.3957	72.5009	76.1638	80.5408	116.3150	122.1071	127.2809	133.4731	137.7983
99	66.5101	69.2299	73.3611	77.0463	81.4492	117.4066	123.2246	128.4209	134.6392	138.9821
100	67.3275	70.0648	74.2219	77.9294	82.3581	118.4973	124.3407	129.5586	135.8008	140.1584

APPENDIX B

CRITICAL VALUES OF SQUARED CORRELATION COEFFICIENT, P-PLOT HYPOTHESIS TEST

Reprinted with permission from: Christensen, R., 1996, *Analysis of Variance, Design, and Regression,* London: Chapman & Hall. Copyright CRC Press, Boca Raton, Florida.

Table B-1 Percentiles of the W' statistic

n	.01	.05	n	.01	.05
5	0.69	0.77	36	0.91	0.940
6	0.70	0.79	38	0.915	0.942
7	0.72	0.81	40	0.918	0.946
8	0.75	0.82	45	0.928	0.951
9	0.75	0.83	50	0.931	0.952
10	0.78	0.83	55	0.938	0.958
11	0.79	0.85	60	0.943	0.961
12	0.79	0.86	65	0.945	0.961
13	0.81	0.870	70	0.953	0.966
14	0.82	0.877	75	0.954	0.968
15	0.82	0.883	80	0.957	0.970
16	0.83	0.886	85	0.958	0.970
17	0.84	0.896	90	0.960	0.972
18	0.85	0.896	95	0.961	0.972
19	0.86	0.902	100	0.962	0.974
20	0.86	0.902	120	0.970	0.978
22	0.87	0.910	140	0.973	0.981
24	0.88	0.915	160	0.976	0.983
26	0.89	0.923	180	0.978	0.985
28	0.89	0.924	200	0.981	0.986
30	0.89	0.928	250	0.984	0.988
32	0.90	0.933	300	0.987	0.991
34	0.91	0.936			

This table was obtained by taking the mean of ten estimates of the percentile each based on a sample of 500 observations. Estimates with standard errors of about .002 or less are reported to three decimal places. The estimates reported with two decimal places have standard errors between about .002 and .008.

APPENDIX C

CRITICAL VALUES OF THE
F DISTRIBUTION

Table C-1	F distribution for selected degrees of freedom and selected cumulative probability levels (one-tail test)							
df	$1 - \alpha$	1	2	3	4	5	6	7	8
1	0.900	39.8634	49.5000	53.5932	55.8329	57.2401	58.2044	58.9059	59.4389
	0.950	161.4475	199.4999	215.7073	224.5832	230.1617	233.9859	236.7681	238.8823
	0.975	647.7870	799.4977	864.1609	899.5812	921.8446	937.1086	948.2143	956.6523
	0.990	4052.1350	4999.4487	5403.2935	5624.5195	5763.5850	5858.9238	5928.2920	5980.9927
	0.995	16210.3525	19999.0605	21614.2441	22499.0957	23055.2617	23436.6035	23714.0508	23924.8438
2	0.900	8.5263	9.0000	9.1618	9.2434	9.2926	9.3255	9.3491	9.3668
	0.950	18.5128	19.0000	19.1643	19.2468	19.2964	19.3295	19.3532	19.3710
	0.975	38.5063	38.9999	39.1654	39.2484	39.2982	39.3314	39.3551	39.3730
	0.990	98.5020	98.9995	99.1657	99.2488	99.2988	99.3321	99.3558	99.3737
	0.995	198.4991	198.9978	199.1642	199.2476	199.2974	199.3307	199.3545	199.3724
3	0.900	5.5383	5.4624	5.3910	5.3426	5.3090	5.2848	5.2657	5.2520
	0.950	10.1280	9.5521	9.2761	9.1169	9.0129	8.9419	8.8856	8.8450
	0.975	17.4434	16.0441	15.4395	15.1029	14.8840	14.7374	14.6274	14.5421
	0.990	34.1161	30.8164	29.4606	28.7088	28.2387	27.9075	27.6723	27.5048
	0.995	55.5515	49.7989	47.4555	46.1732	45.4171	44.8536	44.4028	44.1064
4	0.900	4.5448	4.3246	4.1910	4.1074	4.0507	4.0099	3.9792	3.9549
	0.950	7.7086	6.9443	6.5913	6.3881	6.2559	6.1627	6.0939	6.0412
	0.975	12.2178	10.6491	9.9789	9.6037	9.3636	9.1969	9.0742	8.9801
	0.990	21.1976	17.9999	16.6927	15.9760	15.5202	15.2064	14.9773	14.8014
	0.995	31.3326	26.2841	24.2579	23.1518	22.4552	21.9769	21.6207	21.3526
5	0.900	4.0604	3.7797	3.6196	3.5201	3.4529	3.4044	3.3678	3.3392
	0.950	6.6079	5.7861	5.4097	5.1920	5.0505	4.9504	4.8761	4.8186
	0.975	10.0070	8.4336	7.7637	7.3882	7.1467	6.9781	6.8536	6.7572
	0.990	16.2581	13.2739	12.0596	11.3910	10.9668	10.6727	10.4566	10.2904
	0.995	22.7847	18.3137	16.5302	15.5558	14.9397	14.5140	14.2009	13.9593
6	0.900	3.7759	3.4633	3.2887	3.1808	3.1074	3.0546	3.0144	2.9831
	0.950	5.9874	5.1433	4.7572	4.5337	4.3872	4.2837	4.2065	4.1468
	0.975	8.8131	7.2599	6.5986	6.2274	5.9876	5.8199	5.6956	5.6000
	0.990	13.7450	10.9247	9.7791	9.1488	8.7454	8.4657	8.2597	8.1022
	0.995	18.6349	14.5440	12.9157	12.0284	11.4629	11.0726	10.7863	10.5662

Table C-1		F distribution for selected degrees of freedom and selected cumulative probability levels (one-tail test)							
df	$1 - \alpha$	1	2	3	4	5	6	7	8
7	0.900	3.5894	3.2574	3.0740	2.9605	2.8833	2.8273	2.7850	2.7515
	0.950	5.5914	4.7374	4.3468	4.1204	3.9716	3.8660	3.7870	3.7256
	0.975	8.0727	6.5415	5.8900	5.5228	5.2852	5.1187	4.9952	4.8995
	0.990	12.2464	9.5466	8.4511	7.8469	7.4601	7.1918	6.9927	6.8403
	0.995	16.2355	12.4039	10.8828	10.0510	9.5220	9.1551	8.8855	8.6784
8	0.900	3.4579	3.1131	2.9239	2.8064	2.7264	2.6683	2.6242	2.5893
	0.950	5.3177	4.4590	4.0662	3.8377	3.6874	3.5806	3.5004	3.4382
	0.975	7.5709	6.0595	5.4158	5.0528	4.8174	4.6519	4.5285	4.4330
	0.990	11.2586	8.6491	7.5909	7.0062	6.6315	6.3706	6.1775	6.0291
	0.995	14.6881	11.0424	9.5961	8.8052	8.3020	7.9522	7.6942	7.4963
9	0.900	3.3603	3.0065	2.8129	2.6926	2.6106	2.5508	2.5053	2.4695
	0.950	5.1174	4.2565	3.8625	3.6331	3.4817	3.3737	3.2928	3.2295
	0.975	7.2093	5.7147	5.0781	4.7181	4.4843	4.3197	4.1970	4.1019
	0.990	10.5614	8.0215	6.9920	6.4221	6.0567	5.8015	5.6127	5.4670
	0.995	13.6136	10.1067	8.7172	7.9556	7.4714	7.1341	6.8850	6.6930
10	0.900	3.2850	2.9245	2.7277	2.6054	2.5217	2.4606	2.4140	2.3771
	0.950	4.9646	4.1028	3.7082	3.4781	3.3259	3.2173	3.1354	3.0717
	0.975	6.9367	5.4564	4.8257	4.4682	4.2360	4.0720	3.9497	3.8548
	0.990	10.0443	7.5594	6.5521	5.9942	5.6362	5.3859	5.2003	5.0566
	0.995	12.8264	9.4270	8.0810	7.3429	6.8721	6.5443	6.3028	6.1159
11	0.900	3.2252	2.8595	2.6602	2.5362	2.4512	2.3890	2.3416	2.3040
	0.950	4.8443	3.9823	3.5875	3.3567	3.2039	3.0946	3.0123	2.9481
	0.975	6.7241	5.2559	4.6299	4.2750	4.0441	3.8806	3.7587	3.6638
	0.990	9.6460	7.2057	6.2167	5.6683	5.3162	5.0692	4.8859	4.7445
	0.995	12.2263	8.9122	7.6004	6.8807	6.4219	6.1018	5.8646	5.6824
12	0.900	3.1765	2.8068	2.6055	2.4801	2.3941	2.3310	2.2828	2.2446
	0.950	4.7472	3.8853	3.4904	3.2591	3.1058	2.9962	2.9134	2.8485
	0.975	6.5538	5.0959	4.4742	4.1213	3.8911	3.7283	3.6066	3.5118
	0.990	9.3302	6.9266	5.9525	5.4119	5.0644	4.8205	4.6397	4.4995
	0.995	11.7542	8.5096	7.2255	6.5212	6.0711	5.7569	5.5244	5.3450
13	0.900	3.1362	2.7632	2.5602	2.4337	2.3468	2.2830	2.2341	2.1954
	0.950	4.6672	3.8056	3.4106	3.1791	3.0255	2.9152	2.8320	2.7669
	0.975	6.4143	4.9653	4.3472	3.9960	3.7666	3.6042	3.4828	3.3879
	0.990	9.0738	6.7010	5.7395	5.2053	4.8615	4.6204	4.4411	4.3021
	0.995	11.3735	8.1865	6.9256	6.2334	5.7911	5.4821	5.2531	5.0760
14	0.900	3.1022	2.7265	2.5223	2.3947	2.3070	2.2425	2.1932	2.1539
	0.950	4.6001	3.7389	3.3440	3.1123	2.9582	2.8478	2.7643	2.6986
	0.975	6.2979	4.8567	4.2416	3.8920	3.6635	3.5013	3.3799	3.2854

Table C-1   F distribution for selected degrees of freedom and selected cumulative probability levels (one-tail test)

df	$1 - \alpha$	1	2	3	4	5	6	7	8
	0.990	8.8616	6.5149	5.5638	5.0354	4.6949	4.4559	4.2778	4.1400
	0.995	11.0602	7.9216	6.6805	5.9986	5.5624	5.2573	5.0312	4.8566
15	0.900	3.0732	2.6952	2.4897	2.3614	2.2730	2.2081	2.1581	2.1186
	0.950	4.5431	3.6823	3.2874	3.0556	2.9012	2.7905	2.7066	2.6407
	0.975	6.1995	4.7650	4.1527	3.8043	3.5765	3.4147	3.2933	3.1988
	0.990	8.6831	6.3589	5.4169	4.8933	4.5556	4.3183	4.1415	4.0046
	0.995	10.7980	7.7007	6.4760	5.8028	5.3722	5.0706	4.8472	4.6742
16	0.900	3.0481	2.6682	2.4619	2.3327	2.2438	2.1784	2.1281	2.0880
	0.950	4.4940	3.6337	3.2388	3.0069	2.8525	2.7413	2.6572	2.5910
	0.975	6.1151	4.6867	4.0768	3.7293	3.5022	3.3407	3.2194	3.1249
	0.990	8.5309	6.2262	5.2923	4.7725	4.4375	4.2017	4.0259	3.8897
	0.995	10.5754	7.5138	6.3034	5.6378	5.2117	4.9134	4.6921	4.5205
17	0.900	3.0262	2.6446	2.4374	2.3077	2.2182	2.1524	2.1017	2.0613
	0.950	4.4513	3.5915	3.1967	2.9647	2.8100	2.6987	2.6142	2.5479
	0.975	6.0420	4.6189	4.0111	3.6649	3.4379	3.2766	3.1557	3.0611
	0.990	8.3997	6.1121	5.1850	4.6690	4.3359	4.1016	3.9267	3.7909
	0.995	10.3841	7.3536	6.1557	5.4967	5.0744	4.7791	4.5593	4.3892
18	0.900	3.0070	2.6239	2.4160	2.2857	2.1959	2.1296	2.0785	2.0379
	0.950	4.4139	3.5546	3.1600	2.9277	2.7728	2.6613	2.5767	2.5101
	0.975	5.9780	4.5597	3.9540	3.6083	3.3819	3.2209	3.0999	3.0053
	0.990	8.2854	6.0129	5.0920	4.5791	4.2480	4.0146	3.8407	3.7055
	0.995	10.2181	7.2148	6.0277	5.3747	4.9561	4.6628	4.4449	4.2759
19	0.900	2.9899	2.6056	2.3970	2.2663	2.1759	2.1094	2.0580	2.0171
	0.950	4.3807	3.5219	3.1273	2.8951	2.7401	2.6283	2.5435	2.4768
	0.975	5.9216	4.5075	3.9033	3.5586	3.3328	3.1718	3.$$$	2.9562
	0.990	8.1849	5.9259	5.0104	4.5003	4.1708	3.9385	3.$$$	3.6305
	0.995	10.0725	7.0934	5.9160	5.2681	4.8527	4.5613	4.$$$	4.1771
20	0.900	2.9747	2.5893	2.3800	2.2490	2.1582	2.0913	2.0396	1.9986
	0.950	4.3512	3.4928	3.0985	2.8661	2.7109	2.5989	2.5140	2.4471
	0.975	5.8715	4.4613	3.8588	3.5146	3.2890	3.1283	3.0074	2.9127
	0.990	8.0959	5.8489	4.9382	4.4306	4.1027	3.8715	3.6987	3.5644
	0.995	9.9439	6.9864	5.8176	5.1743	4.7616	4.4721	4.2569	4.0898
25	0.900	2.9177	2.5283	2.3170	2.1843	2.0921	2.0241	1.9713	1.9292
	0.950	4.2417	3.3852	2.9912	2.7586	2.6030	2.4903	2.4047	2.3370
	0.975	5.6864	4.2909	3.6943	3.3530	3.1287	2.9685	2.8479	2.7530
	0.990	7.7698	5.5680	4.6756	4.1773	3.8549	3.6271	3.4568	3.3239
	0.995	9.4753	6.5982	5.4616	4.8351	4.4328	4.1500	3.9394	3.7759

df	$1 - \alpha$	1	2	3	4	5	6	7	8
Table C-1									

Table C-1 F distribution for selected degrees of freedom and selected cumulative probability levels (one-tail test)

df	$1 - \alpha$	1	2	3	4	5	6	7	8
50	0.900	2.8086	2.4119	2.1967	2.0608	1.9660	1.8955	1.8404	1.7963
	0.950	4.0343	3.1826	2.7900	2.5572	2.4004	2.2864	2.1992	2.1300
	0.975	5.3403	3.9749	3.3902	3.0544	2.8326	2.6735	2.5531	2.4579
	0.990	7.1706	5.0566	4.1993	3.7195	3.4077	3.1864	3.0202	2.8901
	0.995	8.6257	5.9016	4.8259	4.2316	3.8486	3.5786	3.3764	3.2189
100	0.900	2.7564	2.3564	2.1394	2.0019	1.9057	1.8339	1.7778	1.7324
	0.950	3.9361	3.0873	2.6955	2.4626	2.3053	2.1906	2.1025	2.0323
	0.975	5.1786	3.8284	3.2496	2.9166	2.6961	2.5374	2.4168	2.3215
	0.990	6.8953	4.8239	3.9837	3.5127	3.2059	2.9877	2.8233	2.6943
	0.995	8.2406	5.5892	4.5424	3.9634	3.5895	3.3252	3.1271	2.9722
∞	0.900	2.7055	2.3026	2.0838	1.9449	1.8473	1.7741	1.7167	1.6702
	0.950	3.8414	2.9957	2.6049	2.3719	2.2141	2.0986	2.0096	1.9384
	0.975	5.0238	3.6889	3.1162	2.7858	2.5665	2.4082	2.2875	2.1918
	0.990	6.6348	4.6052	3.7816	3.3191	3.0172	2.8020	2.6393	2.5112
	0.995	7.8797	5.2983	4.2794	3.7152	3.3499	3.0912	2.8968	2.7444

APPENDIX D

CRITICAL VALUES OF THE
t DISTRIBUTION

Table D-1	t distribution for selected degrees of freedom and selected cumulative probability levels (one-tail test)				
	Cumulative probabilities $P(T \leq t)$				
Degrees of freedom	**0.9000**	**0.9500**	**0.9750**	**0.9900**	**0.9950**
1	3.0777	6.3138	12.7062	31.8206	63.6570
2	1.8856	2.9200	4.3027	6.9646	9.9248
3	1.6378	2.3534	3.1825	4.5407	5.8410
4	1.5332	2.1318	2.7764	3.7470	4.6041
5	1.4759	2.0151	2.5706	3.3649	4.0321
6	1.4398	1.9432	2.4469	3.1427	3.7075
7	1.4149	1.8946	2.3646	2.9980	3.4995
8	1.3968	1.8595	2.3060	2.8965	3.3554
9	1.3830	1.8331	2.2622	2.8215	3.2498
10	1.3722	1.8124	2.2281	2.7638	3.1693
11	1.3634	1.7959	2.2010	2.7181	3.1058
12	1.3562	1.7823	2.1788	2.6810	3.0545
13	1.3502	1.7709	2.1604	2.6503	3.0123
14	1.3450	1.7613	2.1448	2.6245	2.9768
15	1.3406	1.7531	2.1315	2.6025	2.9467
16	1.3368	1.7459	2.1199	2.5835	2.9208
17	1.3334	1.7396	2.1098	2.5669	2.8982
18	1.3304	1.7341	2.1009	2.5524	2.8784
19	1.3278	1.7291	2.0930	2.5395	2.8610
20	1.3253	1.7247	2.0860	2.5280	2.8453
21	1.3232	1.7207	2.0796	2.5176	2.8314
22	1.3213	1.7171	2.0739	2.5083	2.8187
23	1.3194	1.7139	2.0687	2.4999	2.8074
24	1.3178	1.7109	2.0639	2.4921	2.7969
25	1.3164	1.7081	2.0595	2.4851	2.7874
26	1.3150	1.7056	2.0556	2.4786	2.7787
27	1.3137	1.7033	2.0519	2.4727	2.7707
28	1.3125	1.7011	2.0484	2.4671	2.7633
29	1.3114	1.6991	2.0452	2.4620	2.7564
30	1.3104	1.6972	2.0423	2.4573	2.7500
40	1.3031	1.6839	2.0211	2.4232	2.7045
50	1.2987	1.6759	2.0085	2.4033	2.6778
60	1.2958	1.6707	2.0003	2.3902	2.6604
70	1.2938	1.6669	1.9944	2.3808	2.6480
80	1.2922	1.6641	1.9901	2.3739	2.6387
90	1.2910	1.6620	1.9867	2.3685	2.6316
100	1.2901	1.6602	1.9840	2.3642	2.6259
∞	1.2816	1.6449	1.9600	2.3263	2.5758

BIBLIOGRAPHY

Angell, I.O. 1985. *Advanced Graphics with the IBM Personal Computer.* London: Macmillan.

Berthouex, Paul M. and Linfield C. Brown. 1994. *Statistics for Environmental Engineers.* Boca Raton, Florida: CRC Press, 335.

Birkes, David and Yadolah Dodge. 1993. *Alternative Methods of Regression.* New York: Wiley-Interscience, 228.

Brierly, Eric, Paul Sanna, and Anthony Prince, with Timothy Cain, Dejan Jelovic and Robert Stokes. 1998. *Visual Basic 5 How-To.* Corte Madera, California: The Waite Group, 690.

Carr, James R. 1995. *Numerical Analysis for the Geological Sciences.* Englewood Cliffs, New Jersey: Prentice-Hall, 592.

Carr, J. R. and W. B. Benzer. 1991. "On the practice of estimating fractal dimension." *Mathematical Geology,* Vol. 23, No. 7, 945–958.

Carr, J. R. and Korblaah Matanawi. 1999. "Correspondence analysis for principal components transformation of multispectral and hyperspectral digital images." *Photogrammetric Engineering and Remote Sensing,* Vol. 65, No. 8, 909–914.

Chatterjee, S., M. S. Handcock, and J. S. Simonoff. 1995. *A Casebook for a First Course in Statistics and Data Analysis.* New York: John Wiley & Sons.

Chiles, Jean-Paul and Pierre Delfiner. 1999. *Geostatistics: Modeling Spatial Uncertainty.* New York: John Wiley & Sons, 695.

Christensen, Ronald. 1996. *Analysis of Variance, Design and Regression.* London: Chapman & Hall, 587.

Cook, R. Dennis. 1998. *Regression Graphics.* New York: Wiley-Interscience, 349.

Cressie, N. 1991. *Statistics for Spatial Data.* New York: John Wiley and Sons. (Reprinted 1993).

Davis, John C. 1986. *Statistics and Data Analysis in Geology,* 2d ed. New York: John Wiley & Sons.

Deutsch, C. and A. Journel. 1992, 1998. *GSLIB: Geostatistical Software Library and User's Guide.* New York: Oxford University Press.

De Wijs, H. J. 1951. "Statistics of ore distribution. Part I. Frequency distribution of assay values." *Geologie en Mijnbouw,* Vol. 13, No. 11, 365–375.

Englund, E. and A. Sparks. 1988. *GEO-EAS (Geostatistical Environmental Assessment Software) User's Guide,* U.S. EPA 600/4–88/033a.

Englund, E. 1990. "A variance of geostatisticians." *Mathematical Geology,* Vol. 22, No. 4, 417–456.

Galton, F. 1887. "Typical laws of heredity." *Nature,* Vol. 15, 492–495, 512–514, 532–533.

Galton, F. 1886. "Regression towards mediocrity in hereditary stature." *Journal of the Anthropological Institute,* Vol. 15, 246–263.

Goovaerts, P. 1997. *Geostatistics for Natural Resources Characterization.* New York: Oxford University Press, 483.

Greenacre, Michael J. 1984. *Theory and Applications of Correspondence Analysis.* London: Academic Press.

Greenacre, Michael and Jorg Blasius, eds., 1994. *Correspondence Analysis in the Social Sciences.* London: Academic Press, 370.

Haan, C.T. 1977. *Statistical Methods in Hydrology.* Ames, Iowa: Iowa State University Press.

Halvorson, Michael. 1998. *Microsoft Visual Basic 6.0 Step by Step.* Redmond, Washington: Microsoft Press, 632.

Hawkins, D.M. and Cressie, N. 1984. "Robust kriging—a proposal." *Journal of the International Association for Mathematical Geology,* Vol. 16, No. 1, 3–18.

Hurst, H.E. 1951. "Long-term storage capacity of reservoirs." *Transactions of the American Society of Civil Engineers,* Vol. 116, 770–808.

Isaaks, E.H. and R. M. Srivastava. 1989. *An Introduction to Applied Geostatistics.* New York: Oxford University Press, 561.

Jackson, D.A. 1993. "Multivariate analysis of benthic invertebrate communities: the implication of choosing particular data standardizations, measures of association, and ordination methods." *Hydrobiologia,* Vol. 268, No. 1, 9–26.

Jensen, Jerry L., Larry W. Lake, Patrick W. M. Corbett, and David J. Goggin. 1997. *Statistics for Petroleum Engineers and Geoscientists.* Englewood Cliffs, New Jersey: Prentice-Hall, 390.

Journel, A.G. and Ch. J. Huijbregts. 1978. *Mining Geostatistics.* London: Academic Press, 600.

Kermack, K. A. and J.B.S. Haldane. 1950. *Biometrika,* Vol. 37, 30.

Kerman, Mitchell C. and Ronald L. Brown. 2000. *Visual Basic 6.0.* Reading, Massachusetts: Addison-Wesley, 372.

Kitanidis, P.K. 1997. *Introduction to Geostatistics: Applications in Hydrogeology.* New York: Cambridge University Press, 249.

Knudsen, H. P. 1981. *Development of a Conditional Simulation Model of a Coal Deposit.* Tucson, Arizona: University of Arizona, doctoral dissertation.

Kreyszig, Erwin. 1988. Advanced Engineering Mathematics, 6th ed. New York: John Wiley & Sons, 1294 + appendices and index.

Mandelbrot, B.B. 1983. *The Fractal Geometry of Nature.* New York: Freeman.

Magowe, M. and J. R. Carr. 1999. "Relationship between lineaments and ground water occurrence in western Botswana." *Ground Water,* Vol. 37, No. 2, 282–286.

Marcotte, D. 1991. "Cokriging with MATLAB." *Computers & Geosciences,* Vol. 17, No. 9, 1265–1280.

Matheron, G. 1963. Principles of geostatistics. *Economic Geology,* Vol. 58, 1246–1266.

MATLAB *Student Version. 1999. Learning* MATLAB *Version 5.3.* The MathWorks, Inc., Natick, Massachusetts.

McCarn, D.W. and J.R. Carr. 1992. "Effect of numerical precision and equation solution algorithm on computation of kriging weights." *Computers and Geosciences,* Vol. 18, No. 9, 1127–1168.

McCuen, Richard H. 1985. *Statistical Methods for Engineers.* Englewood Cliffs, New Jersey: Prentice-Hall, 439.

Metzger, Stephen and James Carr. 1996. "The role of dust devils in the entrainment of surficial material from the Mars Pathfinder Ares Vallis Landing Site." *NASA Mars Pathfinder Participating Scientist Proposal.*

Metzger, Stephen, James Carr, Jeffrey Johnson, Timothy Parker, and Mark Lemmon. 1999. "Dust devil vortices seen by Mars Pathfinder camera." *Geophysical Research Letters,* Vol. 26, No. 18, September 15, 2781–2784.

Metzger, Stephen, James Carr, Jeffrey Johnson, Timothy Parker, and Mark Lemmon. 2000. "Techniques for identifying dust devils in Mars Pathfinder images." *IEEE Transactions on Geoscience and Remote Sensing,* Vol. 38, No. 2, 870–876.

Middleton, Gerard V. 2000. *Data Analysis in the Earth Sciences Using* MATLAB. Englewood-Cliffs, New Jersey, 260.

Myers, D.E. 1982. "Matrix formulation of cokriging." *Journal of the International Association for Mathematical Geology,* Vol. 14, No. 3, 249–257.

Peitgen, H.O. and D. Saupe, eds. 1988. *The Science of Fractal Images.* New York: Springer Verlag, 312.

Pickover, Clifford A. 1995. *Keys to Infinity.* New York: John Wiley & Sons, 332.

Rao, A. R. and K. H. Hamed. 2000. *Flood Frequency Analysis.* Boca Raton, Florida: CRC Press, 350.

Rodriguez-Iturbe, I., B. Febres de Power, M.B. Sharifi, and K.P. Georgakakos. 1989. "Chaos in Rainfall." *Water Resources Research,* Vol. 25, No. 7, 1667–1675.

Schofield, J.T., J.R. Barnes, D. Crisp, R.M. Haberle, S. Larsen, J.A. Magalhaes, J.R. Murphy, A. Seiff, G. Wilson. 1997. "The MPF atmospheric structure investigation/meteorology [ASI/MET] experiment." *Science*, Vol. 278, 1752–1758.

Schowengerdt, R.A. 1997. *Remote Sensing: Models and Methods for Image Processing*, 2d ed. New York: Academic Press, 522.

Student Edition of MATLAB: version 5. 1997. User's guide/The MathWorks, Inc.: by Duane Hanselman and Bruce Littlefield. Upper Saddle River, New Jersey: Prentice-Hall, 429.

Thomas, P. and P.J. Gierasch. 1985. "Dust devils on Mars." *Science*, Vol. 230, 175–177.

Vincent, R. K. 1997. *Fundamentals of Geological and Environmental Remote Sensing.* Upper Saddle River, New Jersey: Prentice Hall, 366.

Wackernagel, H. 1995. *Multivariate Geostatistics.* Berlin: Springer-Verlag, 256.

York, D. 1966. "Least Squares Fitting of a Straight Line." *Canadian Journal of Physics*, Vol. 44, 1079–1086.

von Mises, Richard. 1957. *Probability, Statistics, and Truth.* New York: Dover Publications, 244.

INDEX

A

Analysis of variance (ANOVA), 61–63
Animation, 233–236
  applied to Mars Pathfinder images, 214–215
Autocorrelation, 85
  spatial autocorrelation, 88–96

B

Bit, def., 182
  relationship to radiometric resolution of digital
    images, 183
Bivariate data analysis, 32–60
  Correlation, 32–60
    correlation coefficient, 42–44
      hypothesis test for, 43–44
  Covariance, 43
  Gauss, Carl (or Karl) F., 32–33
  Method of least squares, 32–58
  Regression, def., 33
  Regression, 33–60
    applications:
      Galton pea data, 33–34
      Geyser eruption data, 38–39
      Mars Pathfinder data, 47–49
      Nevada_Landsat_Data, 49–54
    hypothesis testing for, 40–42
    least absolute deviation, 45–47
    linear, least squares, 32–36
    linear, *type II*—equal error on x and y, 36–38
    M-regression, 46–47
    nonlinear, introduced, 39
      application of, 47–49
    residuals, analysis of, 36, 38–40
    visual methods for, 44–49
    weighted, 58
Byte, def., 182

C

Chi-square (χ^2)
  and correspondence analysis, 73
  for testing conformity of data to a normal
    distribution, 19–22
Coefficient:
  correlation coefficient, 42–44

of kurtosis, 14
of skew, 13
of variation, 11
Cokriging, 132–158
  application to *Nevada_Landsat_Data*, 135–139
  autokrigeability and, 146–151
    relationship to data resolution, 148–151
    with respect to principal components images,
      146–147
  defined, 132–135
  indicator transform used in, 147–148
  paired-sum variograms, 135, 153
  practice of, 139–146
  theory and matrix algebra, 132–139
  undersampling, accommodating, 149
    demonstration using *Nevada_Landsat_Data*,
      149–151
Color compositing, 24-bit bitmap images, 199–200,
  225–227
Conditional simulation, 173–175
Contour line drawing, 225–227, 237–238
Contrast adjustment (of digital images), 181–188
  binary stretch, 186–188
    analogy to indicator transform, 187
  histogram equalization, 185–186
  multiple, linear, 185
  simple, linear, 183–185
  using histograms for, 181–182
Correlation, 32–60
  correlation coefficient, 42–44
    hypothesis test for, 43–44
Correspondence analysis, 71–77
  application to *Nevada_Landsat_Data*, 74–77
Covariance, 43
Cross-variogram (and covariance), 134–137, 158

D

Data Distribution, 15–26
  Exponential distribution, 24–25
  Normal distribution, 16–24
    Log-normal distribtion, 26
  Poisson Distribution, 24
  Probability density functions, 15–26
  Quantiles of, 18
Data, rank order, 18
Data, resolution, 7

SYSTEM REQUIREMENTS

Windows:
Pentium-class processor (or equivalent)
Windows 95 or later
32 MB RAM
2x CD-ROM or better
Standard SVGA monitor, capable of 640x480 resolution, or better

Macintosh:
OS 8.0 or above
32 MB RAM
2x CD-ROM or better
Standard SVGA monitor, capable of 640x480 resolution, or better

WINDOWS AND MACINTOSH COMPATIBILITY

Note: The Visual_Data program is a Windows-based program and not compatible with Macintosh. All data and image files however, are compatible with both Windows and Macintosh.